AF540608

SCIENCE AND TECHNOLOGY OF PIPING DESIGN

SCIENCE AND TECHNOLOGY OF PIPING DESIGN

Manoj Verma

RANDOM PUBLICATIONS

NEW DELHI - 110 002 (INDIA)

Science and Technology of Piping Design

ISBN 978-93-51116-65-3

Published in 2015 in India by

RANDOM PUBLICATIONS

4376-A/4B, Gali Murari Lal, Ansari Road
New Delhi-110 002
Phone: +9111-43580356, 23289044
E-mail: randomexports@gmail.com; sales@randompublications.com; info@randompublications.com
Reprinted 2021

Type Setting by: Friends Media, Delhi-110089
Digitally Printed at: Replika Press Pvt. Ltd.

Preface

A pipe is a tubular section or hollow cylinder, usually but not necessarily of circular cross-section, used mainly to convey substances which can flow — liquids and gases (fluids), slurries, powders, masses of small solids. It can also be used for structural applications; hollow pipe is far stiffer per unit weight than solid members. In common usage the words pipe and tube are usually interchangeable, but in industry and engineering, the terms are uniquely defined. Depending on the applicable standard to which it is manufactured, pipe is generally specified by a nominal diameter with a constant outside diameter (OD) and a schedule that defines the thickness. Tube is most often specified by the OD and wall thickness, but may be specified by any two of OD, inside diameter (ID), and wall thickness. Pipe is generally manufactured to one of several international and national industrial standards. Both "pipe" and "tube" imply a level of rigidity and permanence, whereas a hose (or hosepipe) is usually portable and flexible. Pipe assemblies are almost always constructed with the use of fittings such as elbows, tees, and so on, while tube may be formed or bent into custom configurations. For materials that are inflexible, cannot be formed, or where construction is governed by codes or standards, tube assemblies are also constructed with the use of tube fittings. There are three processes for metallic pipe manufacture. Centrifugal casting of hot alloyed metal is one of the most prominent process. Ductile iron pipes are generally manufactured in such a fashion. Seamless (SMLS) pipe is formed by drawing a solid billet over a piercing rod to create the hollow shell. As the manufacturing process does not include any welding, seamless pipes are perceived to be stronger and more reliable. Historically seamless pipe was regarded as withstanding pressure better than other types, and was often more easily available than welded pipe. There are a number of processes that may be used to produce ERW pipes. Each of these processes leads to coalescence or merging of steel components into pipes. Electric current is passed through the surfaces that have to be welded together; as the components being welded together resist the electric current, heat is generated which forms the weld. Pools of molten metal are formed where the two surfaces are being connected as strong electric current is passed through the metal; these pools of molten metal form the weld that binds the two connected components.

Within industry, piping is a system of pipes used to convey fluids from one location to another. The engineering discipline of piping design studies the efficient transport of fluid. "Piping" sometimes refers to Piping Design, the detailed specification of the physical piping layout within a process plant or commercial building. In earlier days, this was sometimes called Drafting, Technical drawing, Engineering Drawing, and Design but is today commonly performed by Designers who have learned to use automated Computer Aided Drawing / Computer Aided Design (CAD) software.

This book is an essential resource guide for practising engineers and an ideal text for postgraduate students, the book takes a highly practical approach to the design and selection of pipes.

I thank all members of my team who have helped in the preparation of the book. My special thanks go to "Random Publications" who have published the book.

— *Manoj Verma*

Contents

Chapter 1

Introduction

Within industry, piping is a system of pipes used to convey fluids (liquids and gases) from one location to another. The engineering discipline of piping design studies the efficient transport of fluid.

Figure: *Large-scale piping system in an HVAC mechanical room*

Industrial process piping (and accompanying in-line components) can be manufactured from wood, fiberglass, glass, steel, aluminium, plastic, copper, and concrete. The in-line components, known as fittings, valves, and other devices, typically sense and control the pressure, flow rate and temperature of the transmitted fluid, and usually are included in the field of Piping Design (or Piping Engineering). Piping systems are documented in piping and instrumentation diagrams (P&IDs). If necessary, pipes can be cleaned by the tube cleaning process.

"Piping" sometimes refers to Piping Design, the detailed specification of the physical piping layout within a process plant or commercial building. In earlier days, this was sometimes called Drafting, Technical drawing, Engineering Drawing, and Design but is today commonly performed by Designers who have learned to use automated Computer Aided Drawing/Computer Aided Design (CAD) software.

Plumbing is a piping system that most people are familiar with, as it constitutes the form of fluid transportation that is used to provide potable water and fuels to their homes and business. Plumbing pipes also remove waste in the form of sewage, and allow venting of sewage gases to the outdoors. Fire sprinkler systems also use piping, and may transport nonpotable or potable water, or other fire-suppression fluids.

Piping also has many other industrial applications, which are crucial for moving raw and semi-processed fluids for refining into more useful products. Some of the more exotic materials of construction are Inconel, titanium, chrome-moly and various other steel alloys.

Engineering Subfields

Generally, Industrial Piping Engineering has three major subfields:

- Piping Material
- Piping Design
- Stress Analysis.

Stress Analysis

Process piping and power piping are typically checked by Pipe Stress Engineers to verify that the routing, nozzle loads, hangers, and supports are properly placed and selected such that allowable pipe stress is not exceeded under different situation such as sustain, operating, hydro test etc. as per the ASME or any other legislative code and local government standards. It is necessary to evaluate the mechanical behaviour of the piping under regular loads (internal pressure and thermal stresses) as well under occasional and intermittent loading cases such as earthquake, high wind or special vibration, and water hammer. This evaluation is usually performed with the assistance of a specialized (finite element) pipe stress analysis computer προγραμμε such as Caesar II, ROHR2, CAEPIPE and AUTOPIPE.

Wooden Piping History

Early wooden pipes were constructed out of logs that had a large hole bored lengthwise through the center. Later wooden pipes were constructed with staves and hoops similar to wooden barrel construction. Stave pipes have the advantage that they are easily transported as a compact pile of parts on a wagon and then assembled as a hollow structure at the job site. Wooden pipes were especially popular in mountain regions where transport of heavy iron or concrete pipes would have been difficult.

Wooden pipes were easier to maintain than metal, because the wood did not expand or contract with temperature changes as much as metal and so consequently expansion joints and bends were not required. The thickness of wood afforded some insulating properties to the pipes which helped prevent freezing as compared to metal pipes. Wood used for water pipes also does not rot very easily. Electrolysis, that bugbear of many iron pipe systems, doesn't affect wood pipes at all, since wood is a much better electrical insulator. In the Western United States where redwood was used for pipe construction, it was found that redwood had "peculiar properties" that protected it from weathering, acids, insects, and fungus growths. Redwood pipes stayed smooth and clean indefinitely while iron pipe by comparison would rapidly begin to scale and corrode and could eventually plug itself up with the corrosion.

Materials

The material with which a pipe is manufactured often forms as the basis for choosing any pipe. Materials that are used for manufacturing pipes include:

- Carbon Steel (CS)
- Low Temperature Service Carbon Steel (LTCS)
- Stainless Steel (SS)
- Non Ferrous Metals (Inconel, Incoloy, Cupro-nickel, etc.)
- Non Metallic (GRE, PVC, HDPE, tempered glass, etc.).

Standards

There are certain standard codes that need to be followed while designing or manufacturing any piping system. Organizations that promulgate piping standards include:

- ASME-The American Society of Mechanical Engineers
- ASTM-American Society for Testing and Materials
- API-American Petroleum Institute
- AWS-American Welding Society
- AWWA-American Water Works Association
- MSS – Manufacturer's Standardization Society
- ANSI-American National Standard Institute
- NFPA-National Fire Protection Association
- EJMA-Expansion Joint Manufacturers Association.

Black Powder in Gas Pipelines

Black powder is a solid contamination in finished product pipelines. The material may be wet and have a tar-like appearance, or dry and be a very fine powder, sometimes like smoke.

Black powder can cause a range of problems, including product contamination, erosion wear in compressors, instrument and filter clogging and equipment contamination for product consumer, erosion and sealing problems for valves, and flow reduction.

Sources

The source of black powder is far from clear, with several possibilities existing. Black powder could be generated from the following sources:

- Mill scale (iron(II,III) oxide-Fe_3O_4) that comes from pipe manufacturing process through high temperature oxidation of steel. These types of solids are very persistent and strongly adhere to pipe wall and are not easily removed.
- Flash rust ($Fe_2O_3$3, FeOOH) from hydrotest water corrosion.
- Internal pipelines corrosion (microbiological influenced corrosion (MIC) or H_2S reaction with steel.
- Carryover from gas gathering systems.

Black powder may be mechanically mixed or chemically combined with any number of contaminants such as water, liquid hydrocarbons, salts, chlorides, sand, or dirt. Chemical analyses of the material have revealed that it consists mainly of a mixture of iron oxides and iron sulfides.

Sour gas pipelines are always treated with corrosion inhibitors, while sweet gas pipelines are not. In both cases there is a chance of forming black powder at different rates. The rate of inhibition in sour gas lines is designed for normal operation, but there is always a chance of plant upsets that might introduce water to the pipelines. This water is not accounted for and will initiate the formation of black powder. Another source of solids formation in sour gas lines is the mechanical mixing of number of contaminants such as water, liquid hydrocarbons, salts, chlorides, sand, or dirt. If the lines are well inhibited, then the quantity of solids is not significant. Nevertheless, in sweet gas pipelines, which are not inhibited, any water condensation or any plant upset that could introduce water to the line will certainly lead to the formation of black powder.

Methods of Preventing

Cleaning newly installed pipelines to remove mill scale and drying them to remove hydrotest water will delay the formation of black powder. Some water, even at ppm level, exists in the sales gas composition. Therefore, any change in the atmospheric temperature can cause the water vapour to condense in the pipelines. Since water is an important factor in the support of environmental conditions necessary to form black powder, it is expected that the black powder problem in sales gas pipelines is not going to be a one-time occurrence, but it is likely to be cyclic. Accurate water monitoring devices can help to monitor the water dew point at strategic locations on the system to provide an alert when corrective action is required.

Water used to hydrotest pipelines will contribute in the formation of black powder in the long run. Oxygen scavengers are always added to this water; however, no corrosion inhibitors are added. Adding corrosion inhibitors will definitely reduce the chance of forming black powder; however, due to the unavailability of environmentally friendly corrosion inhibitors, the hydrotest water is not treated with corrosion inhibitors. Treating the hydrotest water with corrosion inhibitor will introduce a major environmental concern, especially with this large amount of water.

The most common and historical means of dealing with black powder is to filter it just before it enters a compressor, station, or processing plant. Filters need to be installed in clean pipelines; therefore, newly installed pipelines as well as existing ones need to be cleaned prior and dried completely to filters instillation.

Although sandblasting pipe internals to remove mill scale is one way to remove mill scale, but it will expose the line surface to the atmosphere, and that will increase the corrosion rate during the initial hydrotest. Therefore, chemical or mechanical cleaning, after the initial hydrotesting, might be the best choice and time to remove mill scale and any corrosion products. However and as an alternative, a coating system such as Leighs Paints Pipegard P100 could be utilised after sandblasting. This and other such proprietary coatings not only prevent flash rusting and corrosion, but also promote or do not hinder flow efficiency through the gas line.

The following are recommendations to minimizing solids in newly constructed pipelines:

- Install end caps on the pipe after each day's construction.
- Enhance hydrotest drying operations and dew point monitoring.
- Chemically clean pipelines, using inhibited water, right after the initial hydrotesting.
- Run scrapers as proof of pipeline cleanliness prior to commissioning.

Methods for Removal

Once a line is affected by black powder contamination, its removal becomes a real challenge.

Gas Lines

Black powder in natural gas pipelines becomes extremely hard and thus difficult to remove from the pipeline inner wall. Current scraping technology can reduce the formation of black powder, but there are no documented cases where, once discovered, scraping has been successful in completely removing black powder from a line. Various companies offer methods for removing black powder, including gel scraping, chemical cleaning using diesel/surfactant mixtures/chelants, in-situ chemical cleaning and coating.

LPG Lines

Unlike i.e. wax deposits in crude oil, black powder removed from the pipeline inner wall does not re-dissolve in the fluid (LPG) flow. This feature makes the evacuation of removed black powder out of the line extremely difficult. Especially if the pipeline is long and the amount of black powder significant, accumulation of deposit in front of a scraper pig can easily lead to pipeline blockage. This means that pipelines containing black powder should not be attempted to be cleaned in one pass unless progressive pigging, a by-pass pig, a jet pig, or gel pigs are used.

J. Smart adverts to a sufficient fluid velocity being extremely important for evacuation of removed black powder particles. Consequently, a technology applying enough cleaning force to (1) remove all black powder from the pipeline inner wall and (2) allowing for the required fluid velocity for evacuation is a good choice for cleaning.

Geomagnetically Induced Current, (GIC)

Geomagnetically induced currents (GIC), affecting the normal operation of long electrical conductor systems, are a manifestation at ground level of space weather. During space weather events, electric currents in the magnetosphere and ionosphere experience large variations, which manifest also in the Earth's magnetic field. These variations induce currents (GIC) in conductors operated on the surface of Earth. Electric transmission grids and buried pipelines are common examples of such conductor systems. GIC can cause problems, such as increased corrosion of pipeline steel and damaged high-voltage power transformers. GIC are one possible consequence of geomagnetic storms, which may also affect geophysical exploration surveys and oil and gas drilling operations.

Background

The Earth's magnetic field varies over a wide range of timescales. The longer-term variations, typically occurring over decades to millennia, are predominantly the result of dynamo action in the Earth's core. Geomagnetic variations on timescales of seconds to years also occur, due to dynamic processes in the ionosphere, magnetosphere and heliosphere. These changes are ultimately tied to variations associated with the solar activity (or sunspot) cycle and are manifestations of space weather.

The fact that the geomagnetic field does respond to solar conditions can be useful, for example in investigating Earth structure using magnetotellurics, but it also creates a hazard. This geomagnetic hazard is primarily a risk to technology, at least under the Earth's protective atmospheric blanket.

Risk to Infrastructure

A time-varying magnetic field external to the Earth induces electric currents in the conducting ground. These currents create a secondary (internal) magnetic field. As a consequence of Faraday's law of induction, an electric field at the surface of the Earth is induced associated with time variations of the magnetic field. The surface electric field causes electrical currents, known as geomagnetically induced currents (GIC), to flow in any conducting structure, for example, a power or pipeline grid grounded in the Earth. This electric field, measured in V/km, acts as a voltage source across networks.

Examples of conducting networks are electrical power transmission grids, oil and gas pipelines, undersea communication cables, telephone and telegraph networks and railways. GIC are often described as being quasi direct current (DC), although the variation frequency of GIC is governed by the time variation of the electric field. For GIC to be a hazard to technology, the current has to be of a magnitude and occurrence frequency that makes the equipment susceptible to either immediate or cumulative damage. The size of the GIC in any network is governed by the electrical properties and the topology of the network. The largest magnetospheric-ionospheric current variations, resulting in the largest external magnetic field variations, occur during geomagnetic storms and it is then that the largest GIC occur. Significant variation periods are typically from seconds to about an hour, so the induction process involves the upper mantle and lithosphere.

Since the largest magnetic field variations are observed at higher magnetic latitudes, GIC have been regularly measured in Canadian, Finnish and Scandinavian power grids and pipelines since the 1970s. GIC of tens to hundreds of amperes have been recorded. GIC have also been recorded at mid-latitudes during major storms. There may even be a risk to low latitude areas, especially during a storm commencing suddenly because of the high, short-period rate of change of the field that occurs on the dayside of the Earth.

GIC have been known since the mid-19th century when it was noted that electrical telegraph systems could sometimes run without power during geomagnetic storms, described at the time as operating on the "celestial battery", while at other times they were completely inoperative.

GIC in Power Grids

Modern electric power transmission systems consist of generating plants inter-connected by electrical circuits that operate at fixed transmission voltages controlled at substations. The grid voltages employed are largely dependent on the path length between these substations and 200-700 kV system voltages are common. There is a trend towards higher voltages and lower line resistances to reduce transmission losses over longer and longer path lengths. Low line resistances produce a situation favourable to the flow of GIC. Power transformers have a magnetic circuit that is disrupted by the quasi-DC GIC: the field produced by the GIC offsets the operating point of the magnetic circuit and the transformer may go into half-cycle saturation.

This produces harmonics to the AC waveform, localised heating and leads to high reactive power demands, inefficient power transmission and possible mis-operation of protective measures. Balancing the network in such situations requires significant additional reactive power capacity. The magnitude of GIC that will cause significant problems to transformers varies with transformer type. Modern industry practice is to specify GIC tolerance levels on new transformers.

On 13 March 1989, a severe geomagnetic storm caused the collapse of the Hydro-Québec power grid in a matter of seconds as equipment protective relays tripped in a cascading sequence of events. Six million people were left without power for nine hours, with significant economic loss. Since 1989, power companies in North America, the UK, Northern Europe, and elsewhere have invested in evaluating the GIC risk and in developing mitigation strategies.

GIC risk can, to some extent, be reduced by capacitor blocking systems, maintenance schedule changes, additional on-demand generating capacity, and ultimately, load shedding. These options are expensive and sometimes impractical. The continued growth of high voltage power networks results in higher risk. This is partly due to the increase in the interconnectedness at higher voltages, connections in terms of power transmission to grids in the auroral zone, and grids operating closer to capacity than in the past.

To understand the flow of GIC in power grids and to advise on GIC risk, analysis of the quasi-DC properties of the grid is necessary. This must be coupled with a geophysical model of the Earth that provides the driving surface electric field, determined by combining time-varying ionospheric source fields and a conductivity model of the Earth. Such analyses have been performed for North America, the UK and in Northern Europe.

The complexity of power grids, the source ionospheric current systems and the 3D ground conductivity make an accurate analysis difficult. By being able to analyse major storms and their consequences we can build a picture of the weak spots in a transmission system and run hypothetical event scenarios. Grid management is also aided by space weather forecasts of major geomagnetic storms. This allows for mitigation strategies to be implemented. Solar observations provide a one-to three-day warning of an Earthbound coronal mass ejection (CME), depending on CME speed.

Following this, detection of the solar wind shock that precedes the CME in the solar wind, by spacecraft at the L_1 Lagrangian point, gives a definite 20 to 60 minutes warning of a geomagnetic storm (again depending on local solar wind speed). The magnitude and arrival time of a CME after detection is unknown, although there is much research and model development within the space weather community.

GIC Hazard in Pipelines

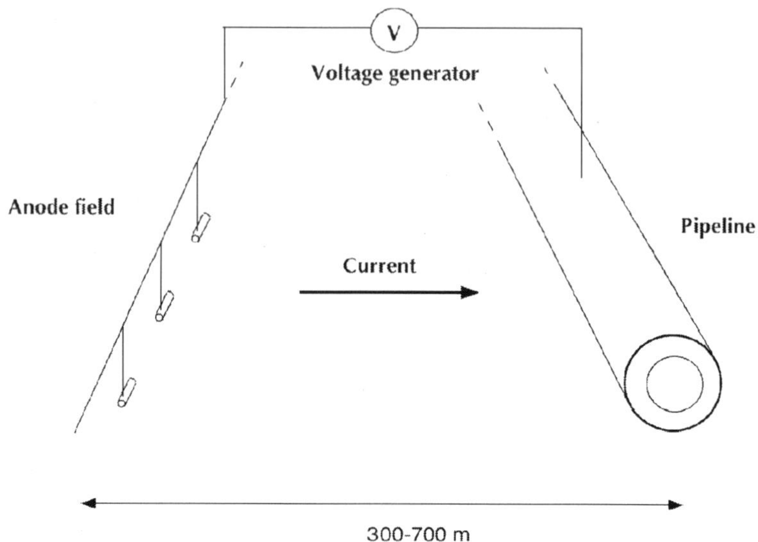

Figure: *Schematic illustration of the cathodic protection system used to protect pipeline from corrosion.*

Major pipeline networks exist at all latitudes and many systems are on a continental scale. Pipeline networks are constructed from steel to contain high-pressure liquid or gas and have corrosion resistant coatings. Weathering and other damage to the pipeline coating can result in the steel being exposed to moist air or to the ground, causing localised corrosion problems. Cathodic protection is used to minimise corrosion by maintaining the steel at a negative potential with respect to the ground. The operating potential is determined from the electrochemical properties of the soil and Earth in the vicinity of the pipeline. The GIC hazard to pipelines is that GIC cause swings in the pipe-to-soil potential, increasing the rate of corrosion during major geomagnetic storms (Gummow, 2002). GIC risk is not a risk of catastrophic failure, but a reduced service life of the pipeline.

Pipeline networks are modelled in a similar manner to power grids, for example through distributed source transmission line models that provide the pipe-to-soil potential at any point along the pipe. These models need to consider complicated pipeline topologies, including bends and branches, as well as electrical insulators (or flanges) that electrically isolate different sections. From a detailed knowledge of the pipeline response to GIC, pipeline engineers can understand the behaviour of the cathodic protection system even during a geomagnetic storm, when pipeline surveying and maintenance may be suspended.

HCNG Dispenser

A HCNG dispenser is a type of fuel dispenser at a filling station that is used to pump HCNG into vehicles.

Principle

A HCNG dispenser is in general combined with a CNG dispenser for natural gas vehicles as both use the same feed stream from the compressed natural gas grid, in addition the hydrogen production method differs per station, some stations use on-site generation where other stations use on-site delivery of dihydrogen to feed the HCNG dispensers. The dispensed blend is 30% H_2 by volume with CNG at 250 bar non-communication fill as HCNG

Locations

HCNG dispensers can be found at Hynor (Norway) Thousand palms and Barstow California Fort Collins Colorado, Dunkerque, Grenoble and Toulouse (France) Goteborg Sweden, Dwarka and Faridabad (Delhi), India and the BC hydrogen highway in Canada.

Hot Tapping

Hot tapping, or pressure tapping, is the method of making a connection to existing piping or pressure vessels without the interruption of emptying that section of pipe or vessel. This means that a pipe or tank can continue to be in operation whilst maintenance or modifications are being done to it. The process is also used to drain off pressurised casing fluids. Hot tapping is also the first procedure in line stopping, where a hole saw is used to make an opening in the pipe, so a line plugging head can be inserted.

Situations in which welding operations are prohibited on equipment which contains:

- Mixtures of gases or vapours within their flammable range or which may become flammable as a result of heat input in welding operations.
- Substances which may undergo reaction or decomposition leading to a dangerous increase in pressure, explosion or attack on metal. In this context, attention is drawn to the possibility that under certain combinations of concentration, temperature and pressure, acetylene, ethylene and other unsaturated hydrocarbons may decompose explosively, initiated by a welding hot spot.

- Oxygen-enriched atmospheres in the presence of hydrocarbons which may be present either in the atmosphere or deposited on the inside surface of the equipment or pipe.
- Compressed air in the presence of hydrocarbons which may be present either in the air or deposited on the inside surfaces of the equipment or pipe.
- Gaseous mixtures in which the partial pressure of hydrogen exceeds 700 kPa gauge, except where evidence from tests has demonstrated that hot-tapping can be done safely.

Based on the above, welding on equipment or pipe which contains hazardous substances or conditions as listed below (even in small quantities) shall not be performed unless positive evidence has been obtained that welding/hot tapping can be applied safely.

Substances:

- Acetylene;
- Acetonitrile;
- Butadiene;
- Caustic soda;*
- Chlorine;
- Compressed air at a pressure in excess of 3000 kPa gauge;
- Ethylene;
- Ethylene oxide;
- Fat/lean DEA/MEA;
- High pressure steam (pressure in excess of 5000 kPa (ga));
- Hydrogen (partial pressure in excess of 700 kPa (ga));
- Hydrogen sulphide;*
- Hydrofluoric acid;
- Oxygen;
- Propene;
- Propene oxide;
- Sulphuric acid;
- Toxic substances.

Constraints based on general hazard in the event of line puncturing during welding, not the welding process. Conditions:

- Vacuum conditions;
- Dissolved hydrogen in the pipe wall (e.g. due to service history);
- Pyrophoric scale deposits.

Hydraulically Activated Pipeline Pigging

Hydraulically activated pipeline pigging (HAPP) is a pigging technology applied for pipeline cleaning. The basic principle is that a pressure drop is created over a by-passable pig held back against a pipeline's fluid flow. The pipeline fluid passing through the pigs cleaning head is accelerated by this pressure drop forming strong cleaning jets. These jets are directed onto the inner wall in front of the pig removing all kinds of deposits. Generally this technology transforms kinetic energy of the pipeline fluid into a locally available differential pressure which in this case is used to create cleaning jets but can also be used otherwise.

Introduction to Pipeline Pigging

Pipeline pigs are devices that are inserted into and travel throughout the length of a pipeline driven by the product flow. They were originally developed to remove deposits which could obstruct or retard flow through a pipeline. Today pigs are used during all phases in the life of a pipeline for many different reasons.

Pigs used today can be divided into three categories :

- Utility Pigs, are used to perform functions such as cleaning, separating, or dewatering.
- In Line Inspection (ILI) Tools, provide information on the condition of the line, as well as the extent and location of any problems.
- Gel Pigs, are used in conjunction with conventional pigs to optimize pipeline dewatering, cleaning, and drying tasks.

Generally for cleaning pigs, the cleaning force applied is the mechanical force between the pipe inner wall and the cleaning pig itself. This force is determined by the pig travel speed as well as by the hardness and shape of the cleaning edge: The faster the pig, the higher the cleaning impact on the deposits but at the same time only the surface of debris is scratched away. Therefore several, sometimes quite many, pig runs are required to clean a pipeline.

Hydraulic activated pigs apply high pressure liquid jets either supplied by high pressure hoses (depended) or made available by the kinetic energy locally available. Depended hydraulic activated pigs are limited in reach due to the hose which needs to be inserted into the pipeline and guides the cleaning head.

HAPP Principle

A hydraulic activated pig consists of three units: a brake unit, a seal unit and the cleaning head.

All units have openings that allow the entire fluid flow through the pipeline to bypass. The brake unit ensures that a hydraulically activated pig is held back against the fluid flow in the pipeline. The fluid pushes against the following seal unit, which channels it into the openings of the cleaning head. Seal unit and cleaning head pose a flow restriction resulting in a pressure difference across the pig. Thus the fluid is accelerated in the cleaning head's nozzles creating extremely powerful liquid jets. These jets are directed onto the pipe inner wall and remove any kind of deposits.

The brake unit ensures that the travel speed of the pig is many times slower than the fluid velocity thus allowing it to entirely remove deposits from the pipe wall before it travels across the cleaned surface. The deposits removed are immediately flushed downstream the pipeline with the main jet of the cleaning head which is directed in the middle of the pipeline. With all deposits removed from the pipe wall and transported downstream by the fluid flow there remains no risk of the pig getting stuck in debris accumulated in front of it.

Hydrogen Piping

Hydrogen piping, in industrial settings, is a system of pipes used to move hydrogen. Due to issues with hydrogen embrittlement, and corrosion, materials for hydrogen pipes must be carefully selected. For metal piping at pressures up to 7,000 psi (48 MPa), high-purity stainless steel piping with a maximum hardness of 80 HRB is preferred.

Hydrogen has an active electron, and therefore behaves somewhat like a Halogen. For this reason, Hydrogen pipes have to resist corrosion. The problem is compounded because hydrogen can easily migrate into the crystal structure of most metals.

Composite pipes are assessed like:

Carbon fiber structure with fiberglass overlay : Carbon-fiber-reinforced polymer or carbon-fiber-reinforced plastic (CFRP or CRP or often simply carbon fiber), is a very strong and light fiber-reinforced polymer which contains carbon fibers. The polymer is most often epoxy, but other polymers, such as polyester, vinyl ester or nylon, are sometimes used. The composite may contain other fibers such as Kevlar, aluminium, glass fibers as well as carbon fiber.

Although it can be relatively expensive, it has many applications in aerospace and automotive fields, as well as in sailboats, and notably finds use in modern bicycles and motorcycles, where its high strength-to-weight ratio and good rigidity is of importance. Improved manufacturing techniques are reducing the costs and time to manufacture, making it increasingly common in small consumer goods as well, such as laptops, tripods, fishing rods, paintball equipment, archery equipment, tent poles, racquet frames, stringed instrument bodies, drum shells, golf clubs, and pool/billiards/snooker cues.

Other terms used to refer to the material: *carbon fiber*, *graphite-reinforced polymer* or *graphite fiber-reinforced polymer* (*GFRP* is less common since it clashes with glass-(fiber)-reinforced polymer). In product advertisements, it is sometimes referred to simply as *graphite fiber* (*graphite fibre*), for short.

Manufacture

The process by which most carbon fiber-reinforced polymer is made varies, depending on the piece being created, the finish (outside gloss) required, and how many of this particular piece are going to be produced. In addition, The choice of matrix can have a profound effect on the properties of the finished composite.

Moulding

One method of producing graphite-epoxy parts is by layering sheets of carbon fiber cloth into a mold in the shape of the final product. The alignment and weave of the cloth fibers is chosen to optimize the strength and stiffness properties of the resulting material. The mold is then filled with epoxy and is heated or air-cured. The resulting part is very corrosion-resistant, stiff, and strong for its weight. Parts used in less critical areas are manufactured by draping cloth over a mold, with epoxy either preimpregnated into the fibers (also known as *pre-preg*) or "painted" over it. High-performance parts using single molds are often vacuum-bagged and/or autoclave-cured, because even small air bubbles in the material will reduce strength.

Vacuum Bagging

For simple pieces of which relatively few copies are needed, (1–2 per day) a vacuum bag can be used. A fiberglass, carbon fiber or aluminium mold is polished and waxed, and has a release agent applied before the fabric and resin are applied, and the vacuum is pulled and set aside to allow the piece to cure (harden).

There are two ways to apply the resin to the fabric in a vacuum mold. One is called a wet layup, where the two-part resin is mixed and applied before being laid in the mold and placed in the bag. The other is a resin induction system, where the dry fabric and mold are placed inside the bag while the vacuum pulls the resin through a small tube into the bag, then through a tube with holes or something similar to evenly spread the resin throughout the fabric.

Wire loom works perfectly for a tube that requires holes inside the bag. Both of these methods of applying resin require hand work to spread the resin evenly for a glossy finish with very small pin-holes. A third method of constructing composite materials is known as a dry layup. Here, the carbon fiber material is already impregnated with resin (prepreg) and is applied to the mold in a similar fashion to adhesive film.

The assembly is then placed in a vacuum to cure. The dry layup method has the least amount of resin waste and can achieve lighter constructions than wet layup. Also, because larger amounts of resin are more difficult to bleed out with wet layup methods, prepreg parts generally have fewer pinholes. Pinhole elimination with minimal resin amounts generally require the use of autoclave pressures to purge the residual gases out.

Compression Molding

A quicker method uses a compression mold. This is a two-piece (male and female) mold usually made out of fiberglass or aluminium that is bolted together with the fabric and resin between the two. The benefit is that, once it is bolted together, it is relatively clean and can be moved around or stored without a vacuum until after curing. However, the molds require a lot of material to hold together through many uses under that pressure.

Filament Winding

For difficult or convoluted shapes, a filament winder can be used to make pieces.

Structure

Many carbon fiber-reinforced polymer parts are created with a single layer of carbon fabric, and filled with fiberglass. A tool called a chopper gun can be used to quickly create these types of parts. Once a thin shell is created out of carbon fiber, the chopper gun is a pneumatic tool that cuts fiberglass from a roll and sprays resin at the same time, so that the fiberglass and resin are mixed on the spot. The resin is either external mix, wherein the hardener and resin are sprayed separately, or internal, where they are mixed internally, which requires cleaning after every use.

The primary element of CFRP is a fibre. From that, an unidirectional sheet is usually created. These can be layered onto each other in a quasi-isotropic layup, e.g. 0, +60,-60 degrees relative to each other.

From the elementary fibre, a bidirectional woven sheet can be created, i.e. a twill with a 2/2 weave.

Properties

The properties of CFRP depends on the layout of the carbon fiber and the proportion of the carbon fibers relative to the polymer.

Automotive Uses

Carbon fiber-reinforced polymer is used extensively in high-end automobile racing. The high cost of carbon fiber is mitigated by the material's unsurpassed strength-to-weight ratio, and low weight is essential for high-performance automobile racing. Racecar manufacturers have also developed methods to give carbon fiber pieces strength in a certain direction, making it strong in a load-bearing direction, but weak in directions where little or no load would be placed on the member. Conversely, manufacturers developed omnidirectional carbon fiber weaves that apply strength in all directions. This type of carbon fiber assembly is most widely used in the "safety cell" monocoque chassis assembly of high-performance racecars.

Many supercars over the past few decades have incorporated CFRP extensively in their manufacture, using it for their monocoque chassis as well as other components.

Until recently, the material has had limited use in mass-produced cars because of the expense involved in terms of materials, equipment, and the relatively limited pool of individuals with expertise in working with it. Recently, several mainstream vehicle manufacturers have started to use CFRP in everyday road cars.

Use of the material has been more readily adopted by low-volume manufacturers who used it primarily for creating body-panels for some of their high-end cars due to its increased strength and decreased weight compared with the glass-reinforced polymer they used for the majority of their products.

Civil Engineering Applications

Carbon fiber reinforced polymer-[CFRP] has over the past two decades become an increasingly notable material used in structural engineering applications. Studied in an academic context as to its potential benefits in construction, it has also proved itself cost-effective in a number of field applications strengthening concrete, masonry, steel, cast iron, and timber structures. Its use in industry can be either for retrofitting to strengthen an existing structure or as an alternative reinforcing (or prestressing material) instead of steel from the outset of a project.

Retrofitting has become the increasingly dominant use of the material in civil engineering, and applications include increasing the load capacity of old structures (such as bridges) that were designed to tolerate far lower service loads than they are experiencing today, seismic retrofitting, and repair of damaged structures. Retrofitting is popular in many instances as the cost of replacing the deficient structure can greatly exceed its strengthening using CFRP.

Applied to reinforced concrete structures for flexure, CFRP typically has a large impact on strength (doubling or more the strength of the section is not uncommon), but only a moderate increase in stiffness (perhaps a 10% increase). This is because the material used in this application is typically very strong (e.g., 3000 MPa ultimate tensile strength, more than 10 times mild steel) but not particularly stiff (150 to 250 GPa, a little less than steel, is typical). As a consequence, only small cross-sectional areas of the material are used. Small areas of very high strength but moderate stiffness material will significantly increase strength, but not stiffness.

CFRP can also be applied to enhance shear strength of reinforced concrete by wrapping fabrics or fibers around the section to be strengthened. Wrapping around sections (such as bridge or building columns) can also enhance the ductility of the section, greatly increasing the resistance to collapse under earthquake loading. Such 'seismic retrofit' is the major application in earthquake-prone areas, since it is much more economic than alternative methods.

If a column is circular (or nearly so) an increase in axial capacity is also achieved by wrapping. In this application, the confinement of the CFRP wrap enhances the compressive strength of the concrete. However, although large increases are achieved in the ultimate collapse load, the concrete will crack at only slightly enhanced load, meaning that this application is only occasionally used.

Specialist ultra-high modulus CFRP (with tensile modulus of 420 GPa or more) is one of the few practical methods of strengthening cast-iron beams. In typical use, it is bonded to the tensile flange of the section, both increasing the stiffness of the section and lowering the neutral axis, thus greatly reducing the maximum tensile stress in the cast iron.

When used as a replacement for steel, CFRP bars could be used to reinforce concrete structures, however the applications are not common.

CFRP could be used as prestressing materials due to their high strength. The advantages of CFRP over steel as a prestressing material, namely its light weight and corrosion resistance, should enable the material to be used for niche applications such as in offshore environments. However, there are practical difficulties in anchorage of carbon fiber strands and applications of this are rare.

In the United States, Prestressed Concrete Cylinder Pipes (PCCP) account for a vast majority of water transmission mains. Due to their large diameters, failures of PCCP are usually catastrophic and affect large populations. Approximately 19,000 miles of PCCP have been installed between 1940 and 2006.

Corrosion in the form of hydrogen embrittlement has been blamed for the gradual deterioration of the prestressing wires in many PCCP lines. Over the past decade, CFRPs have been utilized to internally line PCCP, resulting in a fully structural strengthening system. Inside a PCCP line, the CFRP liner acts as a barrier that controls the level of strain experienced by the steel cylinder in the host pipe. The composite liner enables the steel cylinder to perform within its elastic range, to ensure the pipeline's long-term performance is maintained. CFRP liner designs are based on strain compatibility between the liner and host pipe. CFRP is a more costly material than its counterparts in the construction industry, glass fiber-reinforced polymer (GFRP) and aramid fiber-reinforced polymer (AFRP), though CFRP is, in general, regarded as having superior properties.

Much research continues to be done on using CFRP both for retrofitting and as an alternative to steel as a reinforcing or prestressing material. Cost remains an issue and long-term durability questions still remain. Some are concerned about the brittle nature of CFRP, in contrast to the ductility of steel. Though design codes have been drawn up by institutions such as the American Concrete Institute, there remains some hesitation among the engineering community about implementing these alternative materials. In part, this is due to a lack of standardization and the proprietary nature of the fiber and resin combinations on the market.

Other Applications

The large majority of NHL ice hockey players use carbon-fiber sticks. Carbon-fiber-reinforced polymer has found a lot of use in high-end sports equipment such as racing bicycles. For the same strength, a carbon-fiber frame weighs less than a bicycle tubing of aluminium or steel. The choice of weave can be carefully selected to maximize stiffness. The variety of shapes it can be built into has further increased stiffness and also allowed aerodynamic considerations into tube profiles. Carbon fiber-reinforced polymer frames, forks, handlebars, seatposts, and crank arms are becoming more common on medium-and higher-priced bicycles.

Carbon fiber-reinforced polymer forks are used on most new racing bicycles. Other sporting goods applications include rackets, fishing rods, longboards, and rowing shells.

Much of the fuselage of the new Boeing 787 Dreamliner and Airbus A350 XWB will be composed of CFRP, making the aircraft lighter than a comparable aluminium fuselage, with the added benefit of less maintenance thanks to CFRP's superior fatigue resistance.

Due to its high ratio of strength to weight, CFRP is widely used in micro air vehicles (MAVs). In MAVSTAR Project, the CFRP structures reduce the weight of the MAV significantly. In addition, the high stiffness of the CFRP blades overcome the problem of collision between blades under strong wind.

CFRP has also found application in the construction of high-end audio components such as turntables and loudspeakers, again due to its stiffness.

It is used for parts in a variety of musical instruments, including violin bows, guitar pickguards, and a durable ebony replacement for bagpipe chanters. It is also used to create entire musical instruments such as Blackbird Guitars carbon fiber rider models, Luis and Clark carbon fiber cellos, and Mix carbon fiber mandolins.

In firearms it can substitute for metal, wood, and fiberglass in many areas of a firearm in order to reduce overall weight. However, while it is possible to make the receiver out of synthetic material such as carbon fiber, many of the internal parts are still limited to metal alloys as current reinforced plastics are unsuitable replacements.

Shoe manufacturers use carbon fiber as a shank plate in their basketball sneakers to keep the foot stable. It usually runs the length of the sneaker just above the sole and is left exposed in some areas, usually in the arch of the foot.

CFRP is used, either as standard equipment or in aftermarket parts, in high-performance radio-controlled vehicles and aircraft, i.a. for the main rotor blades of radio controlled helicopters—which should be light and stiff to perform 3D maneuvers.

Fire resistance of polymers or thermoset composites is significantly improved if a thin layer of carbon fibers is molded near the surface—dense, compact layer of carbon fibers efficiently reflects heat.

IBM/Lenovo's ThinkPad laptops and several Sony laptop models use this technology.

Carbon fiber is a popular material to form the handles of high-end knives.

This material is used when manufacturing squash, tennis and badminton racquets.

Carbon-Graphite spars are used on the frames of high-end Sport kites.

In 2006 a company introduced cricket bats with a thin carbon fibre layer on the back which were used in competitive matches by high-profile players (e.g. Ricky Ponting and Michael Hussey). The carbon fibre was claimed to increase the durability of the bats, however they were banned from all first-class matches by the ICC in 2007.

Carbon-fiber is used in the manufacture of high quality arrows for Archery.

End of Useful Life/Recycling

Carbon fiber-reinforced polymers (CFRPs) have a long service lifetime when protected from the sun. When it is time to decommission CFRPs, they cannot be melted down in air like many metals. When free of vinyl (PVC or polyvinyl chloride) and other halogenated polymers, CFRPs can be thermally decomposed via thermal depolymerization in an oxygen-free environment. This can be accomplished in a refinery in a one-step process. Capture and reuse of the carbon and monomers is then possible. CFRPs can also be milled or shredded at low temperature to reclaim the carbon fiber, however this process shortens the fibers dramatically. Just as with downcycled paper, the shortened fibers cause the recycled material to be weaker than the original material. There are still many industrial applications that do not need the strength of full-length carbon fiber reinforcement. For example, chopped reclaimed carbon fiber can be used in consumer electronics, such as laptops. It provides excellent reinforcement of the polymers used even if it lacks the strength-to-weight ratio of an aerospace component.

Despite its high initial strength-to-weight ratio, one structural limitation of CFRP is its lack of a fatigue endurance limit. As such, failure cannot be theoretically ruled out from a high enough number of stress cycles. By contrast, steel and certain other structural metals and alloys do have an estimable fatigue endurance limit. Because of the complex failure modes of such composites, the fatigue failure properties of CFRP are difficult to predict. As a result, when utilizing CFRP for critical cyclic-loading applications, engineers may need to employ considerable strength safety margins to provide suitable component reliability over a sufficiently long service life.

Carbon Nanotube Reinforced Polymer (CNRP)

Carbon nanotube reinforced polymer (CNRP) is several times stronger than CFRP and is being introduced in the Lockheed Martin F-35 Lightning IIý as a structural material.

Perfluoroalkoxy (PFA, MFA)

Perfluoroalkoxy or PFA is a type of fluoropolymer with properties similar to polytetrafluoroethylene (PTFE). It differs from the PTFE resins in that it is melt-processable using conventional injection molding and screw extrusion techniques. PFA was invented by DuPont and is sold under the brandname *Teflon® PFA*. Teflon® is better known as the trade name for PTFE. Other brandnames for granules are Neoflon® PFA from Daikin or Hyflon® PFA from Solvay Solexis.

PFA is very similar in composition to the fluoropolymers PTFE and FEP (fluorinated ethylene-propylene). PFA and FEP both share PTFE's useful properties of low coefficient of friction and non-reactivity, but are more easily formable. PFA is softer than PTFE and melts at 305 °C.

Another version of polytetrafluoroethylene perfluoro methylvinylether, with a different ratio of PTFE and MVE monomers from PFA, is MFA.

Properties

Useful comparison tables of PTFE against FEP, PFA and ETFE can be found on DuPonts website, listing the mechanical, thermal, chemical, electrical and vapour properties of each, side by side.

PFA is similar to FEP in terms of its mechanical properties. These two are both superior to PTFE with regards to their flexibility, making them useful for tubing applications. However, their ability to endure repetitive folding (flex life) is actually lower than PTFEs own. PFA has a higher flex life than FEP.

PFA is preferable to FEP where heat is concerned, but PTFE itself is slightly more resistant to heat than both. PFA is more affected by water absorption and weathering than FEP, but is superior in terms of salt spray resistance.

Perhaps the most impressive feature of PFA is its electrical properties, as it features the dielectric constant of PTFE, and an almost identical dissipation factor, but a dielectric strength that is around four times higher.

ETFE is effectively a high strength engineering version of these three, and features what would likely be considered to be slightly degraded performance when compared with the other three.

Applications

Due to its flexibility, extreme resistance to chemical attack and optical transparency, this material, along with FEP is routinely used for plastic labware and tubing that involves critical or highly corrosive processes. Nalgene is a well known laboratory supplier that makes extensive use of the two materials. Another application are sheet linings for chemical equipment where fabric backed PFA sheets (SYMALIT® PFA). The use of a PFA liner allows to replace expensive metals and alloys like Inconel or Hastelloy with lined carbon steel or lined fibre-reinforced plastic (FRP). Examples are SYMALIT® PFA lined columns, scrubbers, reactors and pipes which stand harsh environments like concentrated HF and HCl and halogens.

Fluorinated Ethylene Propylene (FEP)

Polytetrafluoroethylene (PTFE) is a synthetic fluoropolymer of tetrafluoroethylene that finds numerous applications. PTFE is most well known by the DuPont brand name Teflon.

PTFE is a fluorocarbon solid, as it is a high-molecular-weight compound consisting wholly of carbon and fluorine. PTFE is hydrophobic: neither water nor water-containing substances wet PTFE, as fluorocarbons demonstrate mitigated London dispersion forces due to the high electronegativity of fluorine. PTFE has one of the lowest coefficients of friction against any solid.

PTFE is used as a non-stick coating for pans and other cookware. It is very non-reactive, partly because of the strength of carbon–fluorine bonds, and so it is often used in containers and pipework for reactive and corrosive chemicals. Where used as a lubricant, PTFE reduces friction, wear, and energy consumption of machinery.

It is commonly believed that teflon, like velcro, is a spin-off product from the NASA space projects. However, that is not so, even though both products have been used by NASA.

History

PTFE was accidentally invented by Roy Plunkett of Kinetic Chemicals in New Jersey in 1938. While Plunkett was attempting to make a new CFC refrigerant, the tetrafluoroethylene gas in its pressure bottle stopped flowing before the bottle's weight had dropped to the point signaling "empty."

Since Plunkett was measuring the amount of gas used by weighing the bottle, he became curious as to the source of the weight, and finally resorted to sawing the bottle apart.

Inside, he found it coated with a waxy white material which was oddly slippery. Analysis of the material showed that it was polymerized perfluoroethylene, with the iron from the inside of the container having acted as a catalyst at high pressure.

Kinetic Chemicals patented the new fluorinated plastic (analogous to known polyethylene) in 1941 and registered the Teflon trademark in 1945. An early advanced use was in the Manhattan Project as a material to coat valves and seals in the pipes holding highly reactive uranium hexafluoride at the vast K-25 uranium enrichment plant at Oak Ridge, Tennessee.

DuPont, which founded Kinetic Chemicals in partnership with General Motors, was producing over two million pounds (900 tons) of Teflon per year in Parkersburg, West Virginia, by 1948. In 1954, French engineer Marc Grégoire created the first pan coated with Teflon non-stick resin under the brand name of Tefal after his wife urged him to try the material he had been using on fishing tackle on her cooking pans. In the United States, Kansas City, Missouri resident Marion A. Trozzolo, who had been using the substance on scientific utensils, marketed the first US-made Teflon coated frying pan, "The Happy Pan," in 1961.

Properties

PTFE is a thermoplastic polymer, which is a white solid at room temperature, with a density of about 2.2 g/cm^3. According to DuPont, its melting point is 327 °C (621 °F), but its mechanical properties degrade above 260 °C (500 °F). PTFE gains its properties from the aggregate effect of carbon-fluorine bonds, as do all fluorocarbons.

Property	***Value***
Density	2200 kg/m^3
Melting point	327 °C
Young's modulus	0.5 GPa
Yield strength	23 MPa
Coefficient of friction	0.05–0.10
Dielectric constant	ε=2.1,tan(δ)<5(-4)
Dielectric constant (60 Hz)	ε=2.1,tan(δ)<2(-4)
Dielectric strength (1 MHz)	60 MV/m

The coefficient of friction of plastics is usually measured against polished steel. PTFE's coefficient of friction is 0.05 to 0.10, which is the third-lowest of any known solid material (BAM being the first, with a coefficient of friction of 0.02; diamond-like carbon being second-lowest at 0.05). PTFE's resistance to van der Waals forces means that it is the only known surface to which a gecko cannot stick.

PTFE has excellent dielectric properties. This is especially true at high radio frequencies, making it suitable for use as an insulator in cables and connector assemblies and as a material for printed circuit boards used at microwave frequencies. Combined with its high melting temperature, this makes it the material of choice as a high-performance substitute for the weaker and lower melting point polyethylene that is commonly used in low-cost applications.

Because of its chemical inertness, PTFE cannot be cross-linked like an elastomer. Therefore, it has no "memory" and is subject to creep. This is advantageous when used as a seal, because the material creeps a small amount to conform to the mating surface. However, to keep the seal from creeping too much, fillers are used, which can also improve wear resistance and reduce friction. Sometimes, metal springs apply continuous force to PTFE seals to give good contact, while permitting a beneficially low percentage of creep.

Applications and Uses

Owing to its low friction, it is used for applications where sliding action of parts is needed: plain bearings, gears, slide plates, etc. In these applications, it performs significantly better than nylon and acetal; it is comparable to ultra-high-molecular-weight polyethylene (UHMWPE), although UHMWPE is more resistant to wear than Teflon. For these applications, versions of Teflon with mineral oil or molybdenum disulfide embedded as additional lubricants in its matrix are being manufactured. Its extremely high bulk resistivity makes it an ideal material for fabricating long-life electrets, useful devices that are the electrostatic analogues of magnets.

Gore-Tex is a material incorporating a fluoropolymer membrane with micropores. The roof of the Hubert H. Humphrey Metrodome in Minneapolis is one of the largest applications of Teflon PTFE coatings, using 20 acres (81,000 m^2) of the material in a double-layered, white dome, made with PTFE-coated fiberglass, that gives the stadium its distinctive appearance. The Millennium Dome in London is also made substantially of PTFE.

Powdered PTFE is used in pyrotechnic compositions as oxidizers together with powdered metals such as aluminium and magnesium. Upon ignition, these mixtures form carbonaceous soot and the corresponding metal fluoride, and release large amounts of heat. Hence they are used as infrared decoy flares and igniters for solid-fuel rocket propellants.

In optical radiometry, sheets made from PTFE are used as measuring heads in spectroradiometers and broadband radiometers (e.g., illuminance meters and UV radiometers) due to its capability to diffuse a transmitting light nearly perfectly. Moreover, optical properties of PTFE stay constant over a wide range of wavelengths, from UV up to near infrared. In this region, the relation of its regular transmittance to diffuse transmittance is negligibly small, so light transmitted through a diffuser (PTFE sheet) radiates like Lambert's cosine law. Thus, PTFE enables cosinusoidal angular response for a detector measuring the power of optical radiation at a surface, e.g., in solar irradiance measurements.

PTFE is also used to coat certain types of hardened, armour-piercing bullets, so as to prevent the increased wear on the firearm's rifling that would result from the harder projectile, however it is not the PTFE itself that gives the bullet its armour-piercing property.

High corrosion resistance favours the use of PTFE in laboratory environments as containers, as magnetic stirrer coatings, and as tubing for highly corrosive chemicals such as hydrofluoric acid, which will dissolve glass containers.

PTFE is also widely used as a thread seal tape in plumbing applications, largely replacing paste thread dope.

PTFE membrane filters are among the most efficient used in industrial air filtration applications. Filter coated with a PTFE membrane are often used within a dust collection system to collect particulate matter from air streams in applications involving high temperatures and high particulate loads such as coal-fired power plants, cement production, and steel foundries.

PTFE grafts can be used to bypass stenotic arteries in peripheral vascular disease, if a suitable autologous vein graft is not available.

PTFE can be used to prevent insects climbing up surfaces painted with the material. PTFE is so slippery that insects cannot get a grip and tend to fall off. For example, PTFE is used to prevent ants climbing out of formicaria.

Safety

The pyrolysis of PTFE is detectable at 200 °C (392 °F), and it evolves several fluorocarbon gases and a sublimate. Animal studies indicate that it is unlikely that these products would be generated in amounts significant to health at temperatures below 250 °C (482 °F), although birds are proven to be much more sensitive to these decomposition products.

While PTFE is stable and nontoxic, it begins to deteriorate after the temperature of cookware reaches about 260 °C (500 °F), and decomposes above 350 °C (662 °F). These degradation by-products can be lethal to birds, and can cause flu-like symptoms in humans.

Meat is usually fried between 200 and 230 °C (392 and 446 °F), and most oils will start to smoke before a temperature of 260 °C is reached, but there are at least two cooking oils (refined safflower oil and avocado oil) that have a higher smoke point than 260 °C. Empty cookware can also exceed this temperature upon heating.

PFOA

Perfluorooctanoic acid (PFOA or C8), in the form of the ammonium salt, is used as a surfactant in the emulsion polymerization of PTFE, and has been detected in some PTFE products. The levels that have been measured in nonstick cookware range from not detectable to 75 parts per billion. These are lower than in PTFE products such as thread sealant tape (with 1800 parts per billion of PFOA detected) because nonstick cookware is heated to volatilize PFOA.

A DuPont study on Teflon PTFE did not detect any PFOA above their detection limit of 9 parts per billion, and DuPont says no PFOA is in Teflon cookware. A 2009 USEPA study found levels of PFOA in nonstick cookware ranging from undetected (with a detection limit of 1.5 parts per billion) to 4.3 parts per billion.

DuPont says there should be no measurable amount on a finished pan provided it has been properly cured. While PFOA has been detected in the low parts per billion range in the blood of people, exposure from nonstick cookware is considered insignificant —despite the marketing of other wares. However, at temperatures well above those encountered in cooking, PTFE pyrolysis can form minor amounts of PFOA.

In January 2006, DuPont, the only company that manufactures PFOA in the US, agreed to eliminate releases of the chemical from its manufacturing plants by 2015, but did not commit to completely phasing out its use of the chemical.

In the emulsion polymerization of PTFE, 3M subsidiary Dyneon has developed a replacement emulsifier despite DuPont stating that PFOA is an "essential processing aid". As of August 2008, the EPA's position was that it "has no information that routine use of household or other products using fluoropolymers, such as nonstick cookware or all weather clothing, poses a concern."

Similar Polymers

Other polymers with similar composition are also known by the Teflon trade name:

- Perfluoroalkoxy (PFA)
- Fluorinated ethylene propylene (FEP).

They retain the useful properties of PTFE of low friction and nonreactivity, but are more easily formable. For example, FEP is softer than PTFE and melts at 260 °C (500 °F); it is also highly transparent and resistant to sunlight.

Infrastructure

Water Management Infrastructure

- Drinking water supply, including the system of pipes, storage reservoirs, pumps, valves, filtration and treatment equipment and meters, including buildings and structures to house the equipment, used for the collection, treatment and distribution of drinking water
- Sewage collection, and disposal of waste water
- Drainage systems (storm sewers, ditches, etc.)
- Major irrigation systems (reservoirs, irrigation canals)
- Major flood control systems (dikes, levees, major pumping stations and floodgates)

- Large-scale snow removal, including fleets of salt spreaders, snow-plows, snowblowers, dedicated dump-trucks, sidewalk plows, the dispatching and routing systems for these fleets, as well as fixed assets such as snow dumps, snow chutes, snow melters
- Coastal management, including structures such as seawalls, breakwaters, groynes, floodgates, as well as the use of soft engineering techniques such as beach nourishment, sand dune stabilization and the protection of mangrove forests and coastal wetlands.

Natural Gas Pipeline System in United States

The US natural gas pipeline system is a complex system of pipelines that carries natural gas nationwide and for import and export for use by millions of people daily for their consumer and commercial needs. Across the country, there are more than 210 pipeline systems that total more than 305,000 miles of interstate and intrastate pipelines.

Of the lower 48 US states, those with the most natural gas pipeline running through them are Texas (58,588 miles), Louisiana (18,900), Oklahoma (18,539), Kansas (15,386), Illinois (11,900) and California (11,770). The states with the least natural gas pipeline are Vermont and New Hampshire.

Regulation

The U.S. DOT Office of Pipeline Safety (OPS) administers the national regulatory προγραμμε to assure the safe and environmentally sound transportation of natural gas, liquefied natural gas and hazardous liquids by pipeline. The Federal Energy Regulatory Commission reviews and authorizes the operation of the interstate natural gas pipelines. And, intrastate pipelines that run within one state and do not cross state boundaries are typically regulated by a state government agency. For example in Texas, the Railroad Commission of Texas regulates pipelines, and in Louisiana, it is the Louisiana Department of Natural Resources.

Development

Much of the natural gas pipeline system was constructed in the 1930s and 1940s before many of the then small towns and rural areas across the country were developed into today's larger cities and suburban areas. Learn about the development of the system across the Southwest US, Southeast, Western US, Midwest US, and Northeast US Much of the gas pipeline was made—and continues to be made today – of steel in diameters of 6 inches to up to 48 inches. Many Americans have no idea where or whether a natural gas pipeline runs under their home or office, and since 9/11, for national security purposes, detailed maps of gas pipelines are not available to the general public.

Safety

Over the years, there have been many natural gas explosions involving pipelines in which people have been injured or killed. The most recent was the 2010 San Bruno pipeline explosion that killed at least four people, injured 60 and more victims are still missing. Portions of the San Bruno pipeline had been built in 1956—or 54 years ago. In ideal situations, pipeline inspection gauges or a "PIG" is used to inspect and insure the safe operation of natural gas pipelines. About 63 percent of all natural gas pipelines in the US cannot be properly inspected using a PIG, or automatic robot in the pipes, because the pipelines are either too old or they twist and turn and PIGs cannot operate in them.

Many experts and studies show that the inferior oversight of gas pipelines have led to hundreds of pipeline accidents that have "killed 60 people and injured 230 others in the last five years," according the New York Times. This analysis excludes the casualty figures from the 2010 San Bruno pipeline explosion that killed 7 people and injured more than 50 others.

Chapter 2

Pipeline Components

A pipeline component is a .NET or COM component that implements a set of predefined interfaces for interaction with the BizTalk Messaging Engine. Depending on the functionality of the component, different interfaces must be implemented. This topic discusses these interfaces and some of their methods.

Pipeline Context

All pipeline components can use IPipelineContext methods to access all document processing-specific interfaces. The IPipelineContext interface provides the following functionalities:

- Allows components to retrieve the ambient pipeline and stage settings.
- Allows components to retrieve message and message factories. With these factories, components can create various objects required for the execution of the component.
- Allows components to retrieve the document specifications. A document specification is an XSD schema plus additional annotations.

Base Component

All pipeline components need to implement this interface to provide basic information about the component.

Component

All pipeline components except assemblers and disassemblers implement this interface to get messages from the BizTalk Server engine for processing and to pass processed messages back to the engine.

Execute. Method called by the engine to pass the input message to the component and retrieve the processed message from the component.

Property Bag, IPersistPropertyBag

Pipeline components need to implement IPersistPropertyBag to receive its configuration information. This interface and IPropertyBag are the standard interfaces.

Disassembler Component

A disassembling component is a pipeline component that receives one message on input and produces zero or more messages on output. Disassembling components are used to split interchanges of messages into individual documents.

A disassembler component must implement the methods of the IDisassemblerComponent interface to get messages from BizTalk Server for processing and to pass disassembled documents back to BizTalk Server.

Method	*Description*
Disassemble	Performs the disassembling of the incoming document pInMsg.
GetNext	Gets the next message from the message set that resulted from disassembler execution. Returns NULL if there are no more messages.

Assembler Component

An assembling component is a pipeline component that receives several messages on input and produces one message on output. Assembling components are used to collect individual documents into the message interchange batch. Note In this release of BizTalk Server 2004, assembling functionality is not used, so BizTalk Server always passes one document to the component input.

An assembler component implements the IAssemblerComponent methods that are called by the BizTalk Server engine at run time.

Method	*Description*
AddDocument	Adds the document pInMsg to the list of messages that will be included in the interchange.
Assemble	Builds the interchange from the messages that were added by the previous method. Returns a pointer to the assembled message.

Probe Message

Any pipeline component (general, assembling, or disassembling) can implement IProbeMessage if it requires message probing functionality. A probing component is used in the pipeline stages that have FirstMatch execution mode. In such stages, BizTalk Server gives the message to the component, and the Probe method examines the beginning of the message to determine if the component recognizes the format of the message.

Method	*Description*
Probe	This method takes pInMsg message, and returns True if the format is recognized or False otherwise.

Named Item

This is a helper interface for accessing document schemas from managed and unmanaged code.

Named Item List

This is a helper interface for accessing document schemas from managed and unmanaged code.

Document Spec

Pipeline components can use methods of the IDocumentSpec interface to perform document-specific actions, such as moving content properties to context and back, accessing document schemas, and so on.

Method	*Description*
Doc Type	Returns the type of the current document.
DocSpec Name	Returns the specification name of the current document.
Get Schema Collection	Returns the list of document schemas for the current document.
Get Body Path	Returns the XPath to the node in the document where the body part begins.
Get Distinguished Property Annotation Enumerator	Returns a dictionary enumerator of all distinguished field property annotations.
Get Property Annotation Enumerator	Returns an enumerator of all property annotations.

Component UI

Pipeline components must implement this interface to be used within the Pipeline Designer environment.

Method	*Description*
Icon	Provides the icon that is associated with this component.
Validate	Pipeline Designer calls this method before pipeline compilation to verify that all the configuration properties are set correctly.

The Icon property returns an IntPtr. The following C# example shows how to return an IntPtr.

Developing a Probing Pipeline Component

Any pipeline component (general, assembling, or disassembling) can implement the IProbeMessage interface if it must support message probing functionality. A probing component is used in the pipeline stages that have FirstMatch execution mode. In such stages, the BizTalk Messaging Engine gives the beginning part of the message to the component to determine if the component recognizes the format of the message. If the component recognizes the format, the entire message is given to the component for processing.

The IProbeMessage interface exposes a single method, Probe, which enables the component to check the beginning part of the message. The return value determines whether this component is run. The following steps outline how the BizTalk Messaging Engine runs a stage that requires recognition:

1. If the stage does not contain any components, the stage is not run and the message is given to the subsequent stages for processing.
2. Check if the component implements the IProbeMessage interface. If not, the Messaging Engine invokes the component. The stage processing is done and the message is given to the next stage.
3. The Probe method is invoked. If the return value is True, the component is run. Then the stage processing is done and the message is given to a next stage.
4. The Messaging Engine gets the next component in the stage. If there are no more components and none of the components have been run, it generates an error that pipeline processing has failed. If there are no more components and at least one component has been run, the processing is done.

If a stage does not require recognition (for example, the execution mode is All), the Messaging Engine invokes the component without first querying for the IProbeMessage interface and calling the Probe method.

Reporting Errors from Pipeline Components

Pipeline components report errors in two ways:

- For .NET-based components, by throwing an exception.
- For COM-based components, by setting the ErrorInfo object and returning a failure HRESULT.

Reporting Errors from .NET Pipeline Components

To report an error, a .NET-based pipeline component needs to throw an exception where it reports the error description. To report the name of the component that throws an error, set the Source property of the Exception object.

The Messaging Engine uses the Message and Source properties of the Exception object to report an error. The following message is written to the event log:

> *"There was a failure executing the [receive | send] pipeline: <pipeline name> Source: <Source> [Receive Location | Send Port:] <location | port name> Reason: <Message>."*

Reporting errors from COM pipeline components

To report an error, COM-based pipeline components perform the following actions:

1. The pipeline component sets the IErrorInfo object by calling the SetErrorInfo method.
2. The pipeline component returns a failed HRESULT to the Messaging Engine.

The Messaging Engine uses the GetSource and GetDescription properties of the IErrorInfo object to report an error. If the source is not set, the name of the component is used. If the description is not set or the whole ErrorInfo object is not set, the returned HRESULT is reported instead of the description. The following message is written to the event log:

> *"There was a failure executing the [receive | send] pipeline: <pipeline name> Source: <GetSource> [Receive Location | Send Port:] <location | port name> Reason: <GetDescription or HRESULT>."*

Deploying Pipeline Components

All the NET pipeline component assemblies (native and custom) must be located in the <*installation directory*>\Pipeline Components folder to be executed by the server. If the pipeline with a custom component will be deployed across several servers, the component's binaries must be present in the specified folder on every server.

Custom COM components in the pipeline will also appear in the Toolbox, provided they are registered on the computer as a COM component. Custom .NET pipeline components must be placed into the <*installation directory*>\Pipeline Components folder.

After the binary files are in the correct location, you need to add the component to the Toolbox. For instructions on adding the pipeline component to the Toolbox.

Using the Toolbox

You create a pipeline by dragging components (shapes) from the Toolbox to the design surface. Pipeline Designer helps you assemble valid pipelines by placing certain restrictions on the creation process. You can only select Toolbox components that apply to the pipeline type you are creating. For example, a receive pipeline will show decoders, disassemblers, and validators as valid Toolbox components, while encoders and assemblers will be disabled (dimmed).

In addition, stages can accept only valid components. For example, you cannot drag an encoder component to an Assemble stage. When you drag a component near a valid drop location, an arrow appears, indicating the point where the component can be inserted.

If you drag a valid component onto a stage that is collapsed, pause the mouse over the stage for a few seconds to expand it, and then drop the component into the stage.

Start and End shapes

Start and End shapes appear as bullets on all pipelines. They are provided on the design surface for visual organization only. There is only one of each, and they can neither be added nor deleted. The Start and End shapes do not appear in the Toolbox.

Adding and Removing Components From the Toolbox

Adding a custom component to the Toolbox in Pipeline Designer is similar to the standard Microsoft® Visual Studio® .NET procedure.

To add a pipeline component to the Toolbox :

1. On the Tools menu, click Add/Remove Toolbox Items, or Right-click the Toolbox and then click Add/Remove Items.
2. In the Customize Toolbox dialogue box, click the BizTalk Pipeline Components tab.

   A dialogue box displays the compatible pipeline components. You can navigate through the categories by clicking the tabs at the top of the dialogue box.
3. Select the component you want to add to the Toolbox.
4. Click OK. The component will appear on the BizTalk Pipeline Components tab of the Toolbox.

To remove a pipeline component from the Toolbox

- Open the Customize Toolbox dialogue box (as in the preceding procedure) and clear the component to be removed.

A component remains registered even after it has been removed from the Toolbox.

Creating a New Pipeline

You can add a pipeline template to your project to create a new pipeline.

To Create a New Pipeline

1. In Solution Explorer, select the project in which you want to create the pipeline.
2. On the File menu, click Add New Item.
3. In the Add New Item dialogue box, select a Receive Pipeline or Send Pipeline template by clicking it once.

   Note If you double-click the template, the pipeline will automatically be created with the default name that appears in the Name field.
4. In the Name field, type a name for the pipeline.
5. Click Open.

The new pipeline appears in Solution Explorer. The design surface displays pipeline stages and a default set of components.

Opening a Pipeline

Opening a pipeline file causes Pipeline Designer to appear with the defined pipeline and its stages in the design area. In addition, the Pipeline Designer Toolbox and the Properties window will appear if they are not already open.

To Open a Pipeline

- In Solution Explorer, double-click the pipeline you want to open.
- In Solution Explorer, select the pipeline, and then on the View menu, click Open.

Navigating with the Keyboard

You can navigate through the design surface by using your keyboard instead of your mouse. The following table shows the keys that you can use.

Key	*Effect*
Down Arrow	Moves the cursor to the next connecting line or shape below, including Start and End shapes.
Up Arrow	Moves the cursor to the previous connecting line or shape above, including Start and End shapes.
Page up	Changes the cursor focus to the previous starting shape of a stage. This has the effect of scrolling up the page so that the earlier parts of the pipeline are shown.
Page Down	Changes the cursor focus to the next starting shape of a stage. This has the effect of scrolling down the page so that the next parts of the pipeline are shown.
Home	Changes focus to the Start shape of the pipeline.
End	Changes focus to the End shape of the pipeline.

Creating Pipelines with Pipeline Designer

This section describes how to:

- Create, open, and save pipelines
- Add components to a pipeline
- Use the Schema Collection Property Editor
- Use the Toolbox
- Navigate with the keyboard
- Work with read-only pipeline component properties.

Using Pipeline Designer

Pipeline Designer is a graphical editor, hosted in Microsoft® Visual Studio® .NET, which enables you to create new pipelines; view the pipeline templates included with Microsoft® BizTalk® Server; move pipeline components within a pipeline; and configure pipelines, stages, and pipeline components.

Pipeline Designer uses three key tools of the Visual Studio shell as part of the design experience:

- The Properties window, where most of the characteristics of pipeline objects are viewed and modified.
- The Toolbox, which is used as a source for the design surface.
- The design surface, where components from the Toolbox are dragged and dropped.

Pipeline Designer is integrated with the BizTalk project template to enhance your development experience. After using the project system to create a new BizTalk project, you can use the Add New Item command on the File menu to add a pipeline to your solution.

The Pipeline Designer design surface enables you to draw a graphical representation of a pipeline. The design surface occupies the main section of the Visual Studio .NET window and enables you to edit the pipelines belonging to a BizTalk project. You can navigate between pipelines by clicking the tabs above the design surface.

Each pipeline is composed of stages, with each stage containing one or more components. If there are no components in a stage, a watermark with text indicates that shapes can be inserted from the Toolbox. When the first shape is inserted into a stage, the initial text disappears. The design surface shows the pipeline vertically, running from top (start) to bottom (end). As with other common Microsoft® Windows® programs, you can perform several tasks, such as Open and Save from the File menu.

Using the Biz Talk Project System

You can use the BizTalk® project system to create, organize, and configure BizTalk solutions in the Microsoft® Visual Studio® .NET 2003 environment. The topics and procedures in this section describe how to perform various tasks by using the BizTalk project system. The BizTalk project system uses many of the same project management principles and procedures that you use with other project systems in Visual Studio .NET. This section details common procedures that you might use when creating an application that runs on Microsoft BizTalk Server.

The following figure shows the BizTalk project system design environment with the New Project dialogue box open.

To open Biz Talk Editor

To open BizTalk Editor, you must first create a schema or open a previously created schema.

1. In Solution Explorer, click a schema.
2. On the View menu, click Open, or

   Right-click the schema, and then click Open, or

   Double-click the schema.

   To open BizTalk Mapper

To open BizTalk Mapper, you must first create a map or open a previously created map.

1. In Solution Explorer, click a map.
2. On the View menu, click Open, or

   Right-click the map, and then click Open, or

   Double-click a map.

To open Orchestration Designer

To open Orchestration Designer, you must first create an orchestration or open a previously created orchestration.

1. In Solution Explorer, click an orchestration (.odx).
2. On the View menu, click Open. or
   - Right-click the orchestration, and then click Open. or
   - Double-click the orchestration.

     Orchestration Designer opens.

Chapter 3

Piping and Plumbing Fittings

Fittings are used in pipe and plumbing systems to connect straight pipe or tubing sections, to adapt to different sizes or shapes, and for other purposes, such as regulating or measuring fluid flow. The term *plumbing* is generally used to describe conveyance of water, gas, or liquid waste in ordinary domestic or commercial environments, whereas *piping* is often used to describe high-performance (e.g. high pressure, high flow, high temperature, hazardous materials) conveyance of fluids in specialized applications. The term *tubing* is sometimes used for lighter-weight piping, especially types that are flexible enough to be supplied in coiled form.

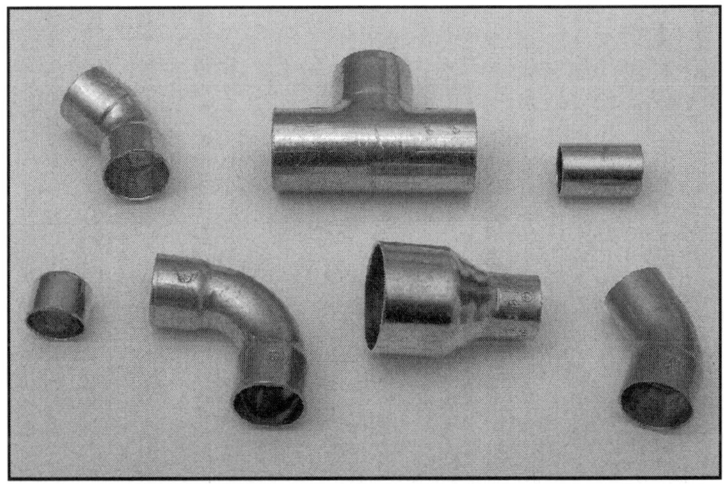

Figure: *Copper fittings for soldered joints*

Fittings (especially uncommon types) require money, time, materials, and tools to install, so they are a non-trivial part of piping and plumbing systems. Valves are technically fittings, but are usually discussed separately.

Standards

There are certain standard codes that need to be followed while designing or manufacturing any piping system. Organizations that promulgate piping standards include:

- ASME-The American Society of Mechanical Engineers
- ASTM-American Society for Testing and Materials
- API-American Petroleum Institute
- AWS-American Welding Society
- AWWA-American Water Works Association
- MSS – Manufacturer's Standardization Society
- ANSI-American National Standard Institute
- NFPA-National Fire Protection Association
- EJMA-Expansion Joint Manufacturers Association.

For example pipes need to conform to the dimensional requirements of :

- ASME B36.10M-Welded and Seamless Wrought Steel Pipe
- ASME B36.19M-Stainless Steel Pipe.

Materials

The material with which a pipe is manufactured often forms as the basis for choosing any pipe. Materials that are used for manufacturing pipes include:

- Carbon Steel(CS)
- Low Temperature Service Carbon Steel (LTCS)
- Stainless Steel (SS)
- Non Ferrous Metals (Inconel, Incoloy, Cupro-nickel etc.)
- Non Metallic (GRE, PVC, HDPE etc.)
- Chrome-molybdenum steel(Alloy steel)-Generally used for high temperature service.

The bodies of fittings for pipe and tubing are most often of the same base material as the pipe or tubing being connected, for example, copper, steel, polyvinyl chloride (PVC), chlorinated polyvinyl chloride (CPVC), or acrylonitrile butadiene styrene (ABS). However, any material that is allowed by the plumbing, health, or building code (as applicable) may be used, but must be compatible with the other materials in the system, the fluids being transported, and the temperatures and pressures inside and outside of the system. For example, brass-or bronze-bodied fittings are common in otherwise copper piping and plumbing systems. Fire hazards, earthquake resistance, and other factors also influence choice of fitting materials.

Gender of Fittings

Piping or tubing are usually (but not always) inserted *into* fittings to make connections. To avoid confusion, connections are conventionally assigned a gender of male or female, respectively abbreviated as "M" or "F".

Common Fittings for Both Piping and Plumbing

While there are hundreds of specialized fittings manufactured, some common types of fittings are used widely in piping and plumbing systems.

Elbow

An elbow is a pipe fitting installed between two lengths of pipe or tubing to allow a change of direction, usually a 90° or 45° angle, though 22.5° elbows are also made. The ends may be machined for butt welding, threaded (usually female), or socketed, etc. When the two ends differ in size, the fitting is called a reducing elbow or reducer elbow.

Figure: *Short radius or regular 45° elbow (copper sweat)*

Elbows are categorized based on various design features as below:

- Long Radius (LR) Elbows – Radius is 1.5 times the pipe diameter
- Short Radius (SR) Elbows – Radius is 1.0 times the pipe diameter

- 90 Deg Elbow – Where change in direction required is 90°
- 45 Deg Elbow – Where change in direction required is 45°

Figure: *Long radius or sweep 90° elbow (copper sweat)*

A 90 degree elbow, also called "90 bends or 90 ells", is a fitting device which is bent in such a way to produce 90 degree change in the direction of flow of the content in the pipe. An Elbow is used to change the direction in piping and is also sometimes called a "quarter bend". A 90 degree elbow attaches readily to plastic, copper, cast iron, steel and lead. It can also attach to rubber with stainless steel clamps. It is available in many materials like silicon, rubber compounds, galvanized steel etc. They connect rigid or IMC circuit together to allow for 90 degree bends. The main application of an elbow (90 degree) is to connect hoses to valves, water pressure pumps, and deck drains. These elbows can be made from tough nylon material or NPT thread. A 45 degree elbow, also called "40 bends or 40 ells", is commonly used in water supply facilities, food industrial pipeline networks, chemical industrial pipeline networks, electronic industrial pipeline networks, air conditioning facility pipeline, agriculture and garden production transporting system, pipeline network for solar energy facility, etc.

Most elbows are available in short radius or long radius types. The short radius elbows have a center-to-end distance equal to the Nominal Pipe Size (NPS) in inches, while the long radius is 1.5 times the NPS in inches. Short elbows are widely available; long elbows are readily available in acrylonitrile butadiene styrene (ABS plastic), polyvinyl chloride (PVC) for DWV, sewage and central vacuums, chlorinated polyvinyl chloride (CPVC) and copper for 1950s to 1960s houses with copper drains.

Coupling

A coupling connects two pipes to each other. If the size of the pipe is not the same, the fitting may be called a reducing coupling or reducer, or an adapter. By convention, the term "expander" is not generally used for a coupler that increases pipe size; instead the term "reducer" is used.

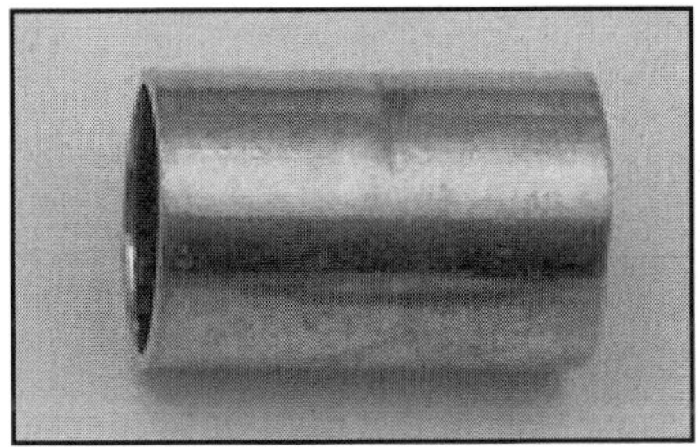

Figure: *Pipe coupling (copper sweat)*

Union

A union is similar to a coupling, except it is designed to allow quick and convenient disconnection of pipes for maintenance or fixture replacement. While a coupling would require either solvent welding, soldering or being able to rotate with all the pipes adjacent as with a threaded coupling, a union provides a simple transition, allowing easy connection or disconnection at any future time. A standard union pipe is made in three parts consisting of a nut, a female end and a male end. When the female and male ends are joined, the nuts then provide the necessary pressure to seal the joint.

Figure: *A combination pipe union and reducer fitting (brass threaded)*

Since the mating ends of the union are interchangeable, changing of a device can be achieved with a minimum loss of time. In addition to a standard union, there exist dielectric unions which are used to separate dissimilar metals (such as copper and galvanized steel) to avoid the damaging effects of galvanic corrosion. When two dissimilar metals are placed in an electrically conductive solution (even tap water is conductive), they will form a battery and generate a voltage by electrolysis.

When the two metals are in contact with each other the current from one metal to the other will cause a movement of ions from one to the other, dissolving one metal and depositing it on the other. A dielectric union breaks the electric current path with a plastic liner between two halves of the union, thus limiting galvanic corrosion.

Reducer

A reducer allows for a change in pipe size to meet hydraulic flow requirements of the system, or to adapt to existing piping of a different size. Reducers are usually concentric but eccentric reducers are used when required to maintain the same top-or bottom-of-pipe level.

Figure: *Reducer fittings, bronze threaded (left) and copper sweat (right)*

Olets

Whenever branch connections are required in size where reducing tees are not available and and/or when the branch connections are of smaller size as compared to header size, o-lets are generally used. The following are few arrangement for o-let connections :

- Flanged Olet
- Socket-Weld & Threaded Olet
- Lateral & Elbow Olets
- Nipple Olet

- Butt-Weld Olet
- Swage Nipples.

Tee

It is the most common pipe fittings. It is available with all female thread sockets, all solvent weld sockets, or with opposed solvent weld sockets and a side outlet with female threads. It is used to either combine or split a fluid flow. Tee is a type of pipe fitting which is T-shaped having two outlets, at 90° to the connection to the main line. It is a short piece of pipe with a lateral outlet. A tee is used for connecting pipes of different diameters or for changing the direction of pipe runs. They are made of various materials and available in various sizes and finishes. They are extensively used in pipeline networks to transport two-phase fluid mixtures. They are categorized as:

- Equal
- Unequal.

Figure: *Pipe tee (copper sweat)*

When the size of the branch is same as header pipes, equal tee is used and when the branch size is less than that of header size, reduced tee will be used. Most common are tees with the same inlet and outlet sizes. Some of the industrial tees are Straight Tee, Reducing Tee, Double Branch Tee, Double Branch Reducing Tee, Conical Tee, Double Branch Conical Tee, Bullhead Tee, Conical Reducing Tee, Double Branch Conical Reducing Tee, Tangential Tee, Double Branch Tangential Tee. The above tees are categorised on the basis of their shapes and structure. They can also be classified on the basis of their application they are required to perform.

Gaskets

Gaskets are the piping item used between flanges for sealing of the joint. Gaskets cannot be reused. Various types of gaskets are available depending upon their construction feature. The following are the type of gaskets commonly used:

- Non-Metallic Gaskets (ASME B 16.21)
- Spiral wound Gaskets (ASME B 16.20)
- Ring Joint Gaskets (ASME B 16.20).

Non-metallic Gaskets are used with flat face or raised face flanges. Spiral wound Gaskets are used with raised face flanges. It is available with inner ring and outer ring which is also known as the cantering ring. Ring Joint Gaskets are used with Ring Type Joint (RTJ) flanges. They are available in octagonal or oval shapes. A very high surface stress is developed between the RTJ gasket and the flange groove when RTJ is bolted up in a flange. This leads to plastic deformation of the gasket. Thus, the hardness of the gasket is kept less than the hardness of the groove hardness to achieve coining. Hence RTJ gaskets are recommended for reuse.

Cross

Cross fittings are also called as 4-way fittings. A cross has one inlet and three outlets, or vice versa. Crosses may have solvent welded socket ends or female threaded ends. They generate a huge amount of stress on the pipe because they have four connection points. A tee is more steady than a cross. If the branch line passes through a tee, the fitting becomes a cross. Crosses are common in fire sprinkler systems, but not in plumbing due to their extra cost as compared to using two tees. The three outlet sizes should be named in order (e.g. left, middle, right; measuring 15-22-15).

Cap

Figure: *Pipe cap (copper sweat)*

A type of pipe fitting, usually liquid or gas tight, which covers the end of a pipe. A cap is used like plug, except that the pipe cap screws on the male thread of a pipe or a nipple. A cap may have a solvent weld socket end or a female threaded end and the other end is closed off. If a solvent weld cap is used to provide for a future connection point, it must be left several inches of pipe before the cap. This is because when the cap is cut off for the future connection, enough pipe is needed to be present to glue a new fitting onto it. In plumbing systems that use threads, the cap has female threads. The industrial caps can be round, square, rectangular, 'U' shaped, 'I' shaped and have a round hand grip or a flat hand grip.

Plug

A plug closes off the end of a pipe. It is similar to a cap but it fits inside the fitting it is mated to. In a threaded iron pipe plumbing system, plugs have male threads. Some of the popular types of plugs are:

- Mechanical pipe plug
- Pneumatic disk pipe plug
- Single size pneumatic all rubber pipe plug
- Multi-size pneumatic pipe plug
- Multi-size flow-through pipe plug
- High pressure pipe plug.

Nipple

A short stub of pipe, usually threaded iron, brass, chlorinated polyvinyl chloride (CPVC) or copper; occasionally just bare copper. A nipple is defined as being a short stub of pipe which has two male ends. Nipples are commonly used for plumbing and hoses, and second as valves for funnels and pipes.

Barb

A barb is used to connect flexible hoses to pipes. The barbed end has a tapered stub with ridges which is inserted into the flexible hose to secure it.

Valves

Valve is equipment designed to stop or regulate flow of any fluid (liquid, gas, condensate, stem, slurry etc.) in its path. Valves are categorized depending on their applications like isolation, throttling and non-return. It is installed in the piping system based on its requirement. Various type of valves are available depending upon the type of construction as follows:

- Gate valve-used for isolation only

- Plug valve-used for isolation only
- Globe valve-used for throttling
- Butterfly valve-used for isolation as well as throttling
- Check valve-used for preventing reverse flow (non-return)
- Diaphragm Valve-used for isolation as well as throttling
- Ball valve-used for isolation only.

Drain-Waste-Vent (DWV) and Related Fittings

Because they operate at low pressure and rely on gravity to move fluids (and often entrained solids), drain-waste-vent systems use fittings designed to be as smooth as possible on their interior surfaces. The fittings may be "belled" or expanded slightly in diameter, or otherwise shaped to accommodate insertion of pipe or tubing, without forming a sharp interior ridge that might catch debris or accumulate buildup of material and cause clogging. The absence of interior snags also makes it much easier to "snake out" or "rod out" a clogged pipe using long flexible tools made for this purpose.

Underground piping systems for landscaping drainage, or disposal of stormwater or groundwater, similarly use gravity flow at low pressure, often with entrained solids. Piping fittings used for these systems bear a strong resemblance to DWV fittings, though often at a larger scale. When high peak flow volumes are involved, the design and construction of these systems are closely inter-related to sewer design. Fittings for central vacuum systems are very similar to DWV fittings, though usually of thinner and lighter construction, since the weight of the materials conveyed through the system is much less. Vacuum system designs share with DWV designs a concern about eliminating internal ridges, burrs, sharp turns, or other obstructions to smooth flow that might cause build-up of material into pipe blockages.

Sweep Elbow

DWV elbows are usually long radius or sweep types, to reduce flow resistance and solids deposition when the direction of flow is changed. A well-designed system will often employ multiple 45° elbows in preference over 90° elbows (even sweep elbows), to reduce flow disruption as much as possible.

Central vacuum system inlet fittings are intentionally designed with a tighter radius of curvature than any other bends in the system. This is done to insure that if any vacuumed debris becomes stuck, it will jam right at the inlet, where it is easiest to discover and to remove.

Closet Flange

The closet flange is the drain pipe flange to which a water closet (toilet) is attached. It is a specialized type of flange connection designed to sit flush with the floor, allowing a standard toilet to be installed above it.

Clean-Outs

Clean-outs are fittings with removable elements that allow access to drains without requiring removal of plumbing fixtures. They are used for allowing an auger or plumber's snake to clean out a plugged drain. Clean-outs should be placed in accessible locations at regular intervals throughout a drainage system, often including outside the building, because clean-out augers have limited length. The minimum requirement is typically at the end of each branch in piping, just ahead of each water closet, at the base of each vertical stack, and both inside and outside the building in the building main drain/sewer. Clean-outs normally have screw-on caps or screw-in plugs. Clean-outs are also known as rodding eyes from the eye-shaped cover plates often used on external versions.

Trap Primers

Trap primers regularly inject water into traps so that "water seals" are maintained, as necessary to keep sewer gases out of buildings. The trap primer must be installed in a readily available place for easy access for adjustments, replacement, and repair. Strictly speaking, a trap primer is a specialized valve, and it is usually connected to a clean water supply, in addition to a DWV system. Because of this dual connection, the design usually must be certified to resist accidental backflow of contaminated water.

Combo-Tee

A combination tee (combo tee) is a tee with a gradually curving center connecting joint. It is used in drain systems to provide a smooth, gradually curving path to reduce the likelihood of clogs, and to ease pushing a plumber's snake through a drain system. The "combo" is a combination of a wye and a 1/8 bend or 45° elbow.

Sanitary Tee

A sanitary tee is a tee with a curved center section designed to minimize the possibility of siphon action that could draw water out of a trap. The center connection is generally connected to the pipe which leads to a trap (the trap arm).

Double Sanitary Tee (Sanitary Cross)

Similar to a cross. This fitting differs from a standard cross in that two of the ports have curved inlets. The fitting has been used in the past for connecting the drains of back-to-back fixtures (such as back-to-back bathroom sinks). Some current codes (including the 2006 UPC) prohibit the use of this fitting for that purpose, instead requiring a special "double fixture fitting".

Wye Fitting

A type of waste fitting tee which has the side inlet pipe entering at a 45° angle. Standard wye is a "Y" shaped fitting device. The branch runs out from the run point at an angle other than 90 degree. It is a fitting with three openings. It is used to create branch lines. A standard wye allows one pipe to be joined to another at a 45 degree angle.

Wyes are similar to tees except that the branch line is angled to reduce friction that could hamper the flow. The connection is typically at a 45-degree angle rather than a 90-degree angle. If a branch turns out at the end to be perpendicular to the through line, the fitting becomes a "tee wye".

PVDF Corrosive Waste Piping Systems utilize wye fittings that feature smooth ID without any irregularities for uninterrupted flow path. These fittings should be able to withstand acids, bases, and solvents, flame-retardant systems withstand intermittent corrosive fluids up to high temperatures. Polypropylene and PVDF Double Wyes and Double Reducing Wyes come in various sizes and are largely used for this purpose. A wye branch allows to split a branch line equally in two directions. The opening sizes can vary for different situations, for instance in situation where a large main line needs to be split into two smaller branches.

The economy wyes are often spot-welded together whereas the industrial wyes have a continuous weld at each seam. A true Wye fitting is also engineered for closed system instrumentation pigging configurations or wherever a smooth pipe branch is required.

Double-Tapped Bushing

A double-tapped bushing is a fitting that has opposing threads on the inside diameter of the bushing.

Hydraulic Fittings

Hydraulic systems use extremely high fluid pressures to create useful work, such as in the hydraulic actuators for powered machinery such as bulldozers and backhoes. Therefore, hydraulic fittings are designed and rated for much greater pressures than those experienced in general piping systems, and they are generally not compatible for use in general plumbing. Fittings are designed and constructed to resist leakage and sudden explosive failure.

Connection Methods

Much of the work of installing a piping or plumbing system involves making leakproof, reliable connections. Depending on the technology used, basic skills may be required, or specialized skills and professional licensure may be required.

Fastener

A fastener is a hardware device that mechanically joins or affixes two or more objects together. The following are the type of fasteners commonly used:

- Stud bolt with nut
- Machine bolt with nut.

Usually the stud bolts are used with full threading and with two heavy hexagonal nuts.

Threaded Pipe

A threaded pipe is a pipe with a screw thread at one or both ends for assembly. Steel pipe is often joined using threaded connections, where tapered threads are cut into the end of the tubing segment, sealant is applied in the form of thread sealing compound or thread seal tape (also known as PTFE or Teflon tape), and it is then threaded into a corresponding threaded fitting using a pipe wrench.

Figure: *MIP pipe and FIP elbow fitting (steel threaded)*

Threaded steel pipe is still widely used in many homes and businesses to convey natural gas or propane fuel, and is a popular choice in fire sprinkler systems due to its high heat resistance. Threaded brass pipe was once used in a similar fashion, and was considered superior to steel for carrying drinking water, but is now effectively obsolete.

Assembling threaded steel pipe takes some skill, plus careful planning to allow lengths of pipe to be screwed together in proper sequence. Most threaded pipe systems require occasional use of pipe union fittings to allow final assembly.

Solvent Welding

A solvent is applied to PVC, CPVC, ABS, or other plastic piping, to partially dissolve and fuse the adjacent surfaces of piping and fitting. Solvent welding is usually used with a sleeve-type joint, to connect pipe and fittings made of the same (or closely compatible) material.

Unlike regular welding of metals, solvent welding is relatively easy to perform, although care is still needed to produce reliable joints. Solvents typically used for plastics are usually toxic, may be carcinogenic, and may also be flammable, requiring adequate ventilation.

Soldering

To make a solder connection, a chemical flux is applied to the inner sleeve of a sleeve type joint, and the pipe is inserted. The joint is then heated using a propane torch or MAPP gas torch, solder is applied to the heated joint, and the melted solder is drawn into the joint by capillary action as the flux vaporizes. Sweating is an alternate term sometimes used to describe soldering of pipe joints.

In situations where many connections must be made in a short period of time (such as plumbing of a new building), solder offers much quicker and much less expensive joinery than compression or flare fittings. A degree of skill is needed to rapidly make large numbers of reliable soldered joints. If flux residues are thoroughly cleaned, soldering can produce a long-lasting connection at low cost. However, the use of open flames for heating joints can present fire and health hazards to occupants of a building being worked on, and requires adequate ventilation.

Brazing

Brazing is closely related to soldering, but uses harder materials and higher temperatures.

Welding

Welding of metals differs from soldering and brazing, in that the connection is made without adding a special low-melting-point material (e.g. solder) to complete a joint. Instead, the material of the pipe or tubing is itself partially melted in a carefully controlled manner, and the fitting and piping are directly fused together. Generally this requires that the piping and the fitting be made of the same (or closely compatible) material. Pipe welding is a specialized skill, and is often performed by specially licensed workers who are tested periodically for the quality of their work. For ultra-critical applications, every joint is tested using non-destructive testing methods.

Properly welded and inspected joints are considered to be very reliable, robust, and long-lasting. Proper ventilation is essential to remove dangerous metallic fumes from welding operations. Because of the expensive skilled labour required, welded pipe joints are usually restricted to high-performance applications, such as in shipbuilding, chemical reactors, and nuclear reactors.

Compression Fittings

Compression fittings consist of a tapered concave conical seat, a hollow barrel-shaped compression ring, and a compression nut which is threaded onto the body of the fitting and tightened to make a leakproof connection. Fittings are typically made of brass or plastic, but stainless steel or other materials may be used.

Figure: *Compression connectors on an isolating valve. The red metal is a copper compression ring.*

Compression connections do not typically have the long life that sweat connections offer, but are advantageous in many cases because they are easy to make using basic tools. A disadvantage of compression connections is that they take longer to make than sweated joints, and sometimes require retightening over time to stop leaks.

Flare Fittings

Flared connectors should not be confused with compression connectors, with which they are generally not interchangeable. Flared connectors lack a compression ring, but do use a threaded nut. A special flaring tool is used to enlarge tubing into a tapered "bellmouth" shape that matches the tapered projecting conical shape of the flare fitting. The flare nut, which has previously been installed over the tubing, is then tightened onto the fitting. Fittings are typically made of brass or plastic, but stainless steel or other materials may be used.

Flare connections are a labour intensive method of making connections, but are quite reliable over the course of many years. Flared fittings are sometimes thought to be more secure against leaks and sudden failures, and are often preferred for safety-critical connections, such as in hydraulic brake systems.

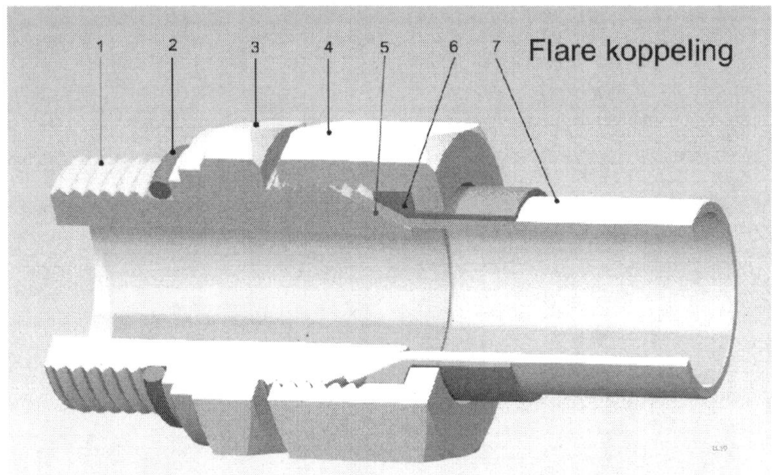

Figure: *Flare connection: 1 Screw thread, 2 O-ring, 3 Body, 4 Nut, 5 Seal interface, 6 Support ring (sleeve), 7 Flared tubing.*

Flange Fittings

Flanges are generally used when there is connection to valves, in-line instruments and/or connection to equipment nozzles is required. Flange fittings generally involve pressing two surfaces to be joined tightly together, by means of threaded bolts, wedges, clamps, or other means of applying high compressive forces. Often, a gasket, packing, or an O-ring is installed between the flanges to prevent leakage, but it is sometimes possible to use only a special grease, or nothing at all, if the mating surfaces are precisely formed. Flanges are designed to the following pressure ratings: 150lb, 300lb, 400lb, 600lb, 900lb,1500lb and 2500lb or 10 Bar, 15Bar, 25Bar, 40Bar, 64Bar, 100Bar and 150Bar. Various types of flanges are available depending upon the type of their constructional features. The following are types of flanges generally used in piping. These flanges are available with different facing like raised face, flat face, ring joint face etc.

- Orifice
- Threaded
- Slip-On
- Blind
- Weld Neck
- Socket
- Lap Joint
- Reducing.

Slip On flanges are slipped over the pipe and then welded from both inside and outside to provide sufficient strength and prevent leakage. This flange is used instead of weld necks by many users because of its lower cost and also the fact that it requires less accuracy when cutting pipe to length.

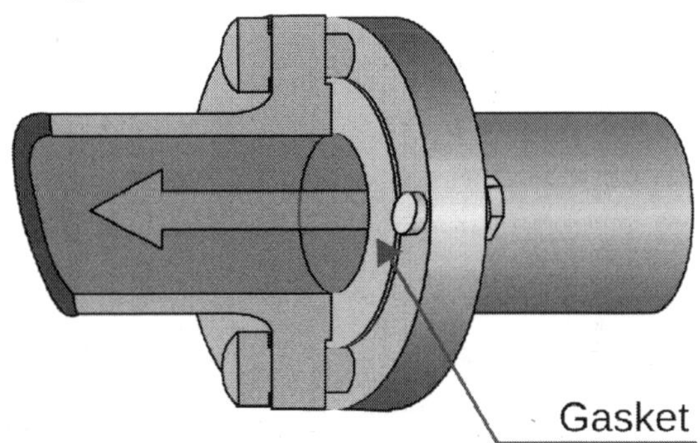

Figure: *Flange connection, using a gasket*

Blind flanges do not have a bore and is used to shut off a piping system or vassal opening. Its design permits easy access to vassal or piping system for inspection purpose. It can be supplied with or without hubs at the manufacturer's option.

Weld Neck Flanges are designed to be joined to a piping system by butt welding. They are expensive because of its long neck, but is preferred for high stresses to the pipe, reducing stress applications. The neck, or hub, transmits stress concentration at the base of the flange. The gradual transition of thickness from the base of the hub to the wall thickness at the butt weld provides important reinforcement of the flange. Troublence and erosion are reduced due to the matching bore size of the pipe and flange.

Socket Flanges are similar to a slip on flanges in outline, but the bore is counter-bored to accept pipe. The diameter of the remaining bore is same as the inside diameter of the pipe. A fillet weld around the hub of the flange attaches the flange to the pipe. An optional interval weld may be applied in high stress applications. Its biggest use is in high pressure system such as hydraulic and steam lines. Lap Joint flange is again similar to a slip flange, but it has radius at the intersection of the bore and the flange face to accommodate a lap stub end. The face on the stub end forms the gasket face of the flange. Its applications are where sections of piping systems need to be dismantled quickly and easily for inspection or replacement.

Flanged connections tend to be more bulky than other connections, but can perform well in demanding applications, such as large water mains and hydroelectric power systems.

Mechanical Fittings

Manufacturers such as Victaulic or Grinnell produce special sleeve clamp fittings that are increasingly replacing classic flanged connections. They typically attach to the end of a pipe segment by using circumferential grooves pressed (or cut, in older designs) around the end of the pipe to be joined. Mechanical connectors are widely used on larger steel pipes, but can also be used with other materials.

Figure: *Multiple mechanical sleeve clamps connecting piping*

The chief advantage of these newer connectors is that they can be installed in the field after cutting the pipe to length, which is much faster than traditional flanged connections, which must be factory-welded or field-welded to pipe segments.

Crimped or Pressed Fittings

Crimped or pressed connections use special fittings which are permanently attached to tubing with a powered crimper. The special fittings, manufactured with sealant already inside, slide over the tubing to be connected. High pressure is used to deform the fitting and compress the sealant against the inner tubing, creating a leakproof seal.

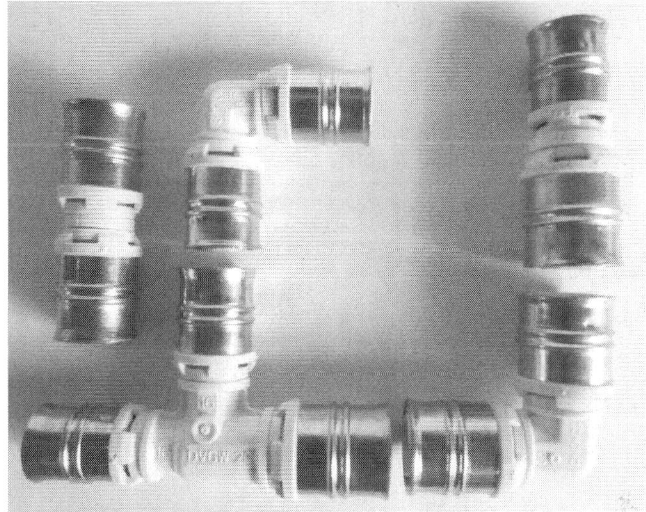

Figure: *Fittings for crimping*

The advantages of this method are that it should last as long as the tubing, it takes less time to complete than other methods, it is cleaner in both appearance and the materials used to make the connection, and no open flame is used during the connection process. The disadvantages are that the fittings used are harder to find and cost significantly more than sweat type fittings.

Tap (valve)

A tap (also called spigot and faucet in the U.S.) is a valve controlling release of liquids or gas. In the British Isles and most of the Commonwealth, the word is used for any everyday type of valve, particularly the fittings that control water supply to bathtubs and sinks. In the U.S., the term "tap" is more often used for beer taps, cut-in connections, or wiretapping. "Spigot" or "faucet" are more often used to refer to water valves, although this sense of "tap" is not uncommon, and the term "tap water" is the standard name for water from the faucet.

Figure: *Indoor taps are commonly found in the bathroom or kitchen. This tap is a single-handle, double-spout tap (one spout for hot, one spout for cold).*

Water Taps

Figure: *Old metal taps were often part of decoration as this one in Fužine castle backyard, Slovenia.*

The physical characteristic which differentiates a spigot from other valves is the lack of any type of a mechanical thread or fastener on the outlet. Water for baths, sinks and basins can be provided by separate hot and cold taps; this arrangement is common in older installations, particularly in public washrooms/lavatories and utility rooms/laundries. In kitchens and bathrooms mixer taps are commonly used. In this case, hot and cold water from the two valves is mixed together before reaching the outlet, allowing the water to emerge at any temperature between that of the hot and cold water supplies. Mixer taps were invented by Thomas Campbell of Saint John, New Brunswick and patented in 1880.

For baths and showers, mixer taps frequently incorporate some sort of *pressure balancing* feature so that the hot/cold mixture ratio will not be affected by transient changes in the pressure of one or the other of the supplies.

This helps avoid scalding or uncomfortable chilling as other water loads occur (such as the flushing of a toilet). Rather than two separate valves, mixer taps frequently use a single, more complex, valve controlled by a single handle (single handle mixer). The handle moves up and down to control the amount of water flow and from side to side to control the temperature of the water. Especially for baths and showers, the latest designs do this using a built in thermostat. These are known as thermostatic mixing valves, or TMVs, and can be mechanical or electronic. There are also faucets with colour LEDs to show the temperature of the water.

Figure: *North American shower tap. Lower lever controls water exit; right: to shower ("SHR"), left: to bathtub ("TUB"), middle: no water. Middle lever temperature controls: turn counterclockwise to augment water flow, turn further to increase temperature.*

If separate taps are fitted, it may not be immediately clear which tap is hot and which is cold. The hot tap generally has a red indicator while the cold tap generally has a blue or green indicator. In the United

States, the taps are frequently also labelled with an "H" or "C". Note that in countries with Romance languages, the letters "C" for hot and "F" for cold are used (from French "chaud"/Italian "caldo" (hot) and French "froid"/Italian "freddo" (cold)). This can create confusion when English speakers visit these countries or vice versa. Mixer taps may have a red-blue stripe or arrows indicating which side will give hot and which cold.

Figure: *Hot is on the left in many countries.*

In most countries, there is a standard arrangement of hot/cold taps. For example, in the United States and many other countries, the hot tap is on the left by building code requirements. Many installations exist where this standard has been ignored (called "crossed connections" by plumbers). Mis-assembly of some single-valve mixer taps will exchange hot and cold even if the fixture has been plumbed correctly.

Most handles on residential homes are connected to the valve shaft and fastened down with a screw. Although on most commercial and industrial applications they are fitted with a removable key called a "loose key", "water key", or "sillcock key", which has a square peg and a square ended key to turn off and on the water. You can also take off the "loose key" to prevent vandals from turning on the water. In older buildings before the "loose key" was invented for some landlords or caretakers to take off the handle of a residential tap, which had teeth that would meet up with the gears on the valve shaft. This Teeth and cog system is still used on most modern faucets. Although most of the time a "loose key" is on industrial and commercial applications sometimes you may see a "loose key" on homes by the seashore to prevent passers-by from washing the sand off their feet.

Beer Taps

Figure: *A gravity cask tap.*

While in other contexts, depending on location, a "tap" may be a "faucet", "valve" or "spigot", the use of *"tap"* for beer is almost universal. This may be because the word was originally coined for the wooden valve in traditional barrels. Draft beer dispensed with a valve is said to be "on tap" (as an idiom). A "beer tap" now may be one of several items:

Pressure-Dispense Bar Tap: Almost universally in modern times, bulk beer is supplied in kegs that are served with the aid of external pressure. In a normal bar dispense system, this pressure comes from a cylinder of carbon dioxide in North America or nitrogen in the UK which forces the beer out of the keg and up a narrow tube to the bar. At the end of this tube is a valve built into a fixture (usually somewhat decorative) on the bar. This is the beer tap, and opening it with a small lever causes beer, pushed by the gas from the cylinder, to flow into the glass.

Portable Keg Tap: Sometimes, beer kegs designed to be connected to the above system are instead used on their own, perhaps at a party or outdoor event. In this case, a self-contained portable tap is required that allows beer to be served straight from the keg. Because the keg system uses pressure to force the beer up and out of the keg, these taps must have a means of supplying it. The typical "picnic tap" uses a hand pump to push air into the keg; this will cause the beer to spoil faster but is perfectly acceptable when it will be consumed in a short time. Portable taps with small CO_2 cylinders are also available.

Cask Beer Tap: Beers brewed and served in the traditional way (typically cask ale) do not use artificial gas. Taps for cask beer are simple on-off valves that are hammered into the end of the cask. When beer is served directly from the cask ("by gravity"), as at beer festivals and some pubs, it simply flows out of the tap and into the glass. When the cask is stored in the cellar and served from the bar, as in most pubs, the beer line is screwed onto the tap and the beer is sucked through it by a hand-operated low-pressure pump on the bar. The taps used are the same, and in beer-line setups the first pint is often poured from the cask as for "gravity", for tasting, before the line is connected. Cask beer taps can be brass (now discouraged for fear of lead contamination), stainless steel (good, but expensive), plastic (acceptable, and cheaper), and wood (to be avoided if possible).

Gas Taps

Although a gas tap may be a valve that releases any gas, the word is most commonly used to refer to taps that control the flow of fuel gas (natural gas or, historically, coal gas, syngas, etc.) in the home (for gas fires or other appliances) or in laboratories (for Bunsen burners).

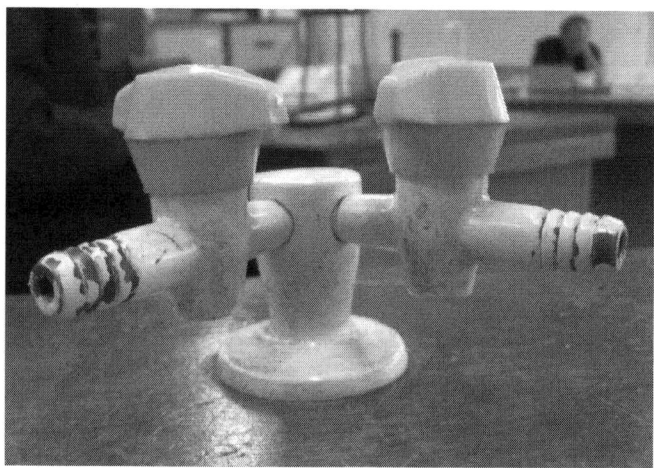

Figure: *Gas taps*

Physics of Taps

Most water and gas taps have adjustable flow. Turning the knob or working the lever sets the flow rate by adjusting the size of an opening in the valve assembly, giving rise to choked flow through the narrow opening in the valve. The choked flow rate is independent of the viscosity or temperature of the fluid or gas in the pipe, and depends only weakly on the supply pressure, so that flow rate is stable at a given setting. At intermediate flow settings the pressure at the valve restriction drops nearly to zero from the venturi effect; in water taps, this causes the water to boil momentarily at room temperature as it passes through the restriction. Bubbles of cool water vapour form and collapse at the restriction, causing the familiar hissing sound. At very low flow settings, the viscosity of the water becomes important and the pressure drop (and hissing noise) vanish; at full flow settings, parasitic drag in the pipes becomes important and the water again becomes quiet.

One reason that most beer taps are not designed for adjustable flow is that the beer itself is damaged by the pressure drop in a choked-flow valve: holding a beer tap partially open causes the beer to foam vigorously, ruining the pour.

Tap Mechanisms

The first screw-down tap mechanism was patented and manufactured by the Rotherham brass founders Guest and Chrimes in 1845. Most older taps use a soft rubber or neoprene washer which is screwed down onto a valve seat in order to stop the flow. This is called a "globe valve" in engineering and, while it gives a leak-proof seal and good fine adjustment of flow, both the rubber washer and the valve seat are subject to wear (and for the seat, also corrosion) over time, so that eventually no tight seal is formed in the closed position, resulting in a leaking tap. The washer can be replaced and the valve seat resurfaced (at least a few times), but globe valves are never maintenance-free.

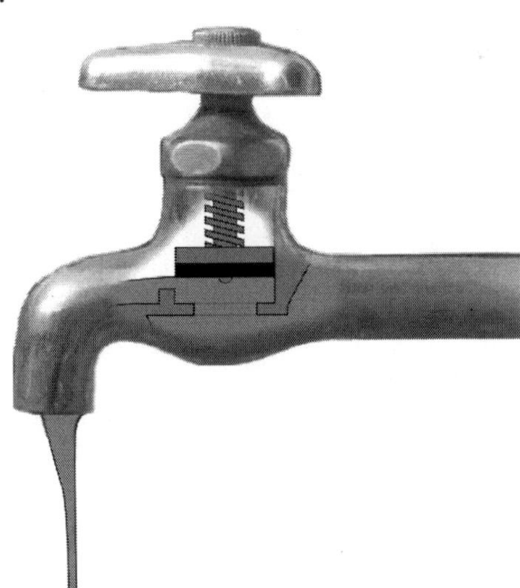

Figure: *Tap mechanism*

Also, the tortuous S-shaped path the water is forced to follow offers a significant obstruction to the flow. For high pressure domestic water systems this does not matter, but for low pressure systems where flow rate is important, such as a shower fed by a storage tank, a "stop tap" or, in engineering terms, a "gate valve" is preferred.

Gate valves use a metal disc the same diameter as the pipe which is screwed into place perpendicularly to the flow, cutting it off. There is no resistance to flow when the tap is fully open, but this type of tap rarely gives a perfect seal when closed. In the UK this type of tap normally has a wheel-shaped handle rather than a crutch or capstan handle.

Cone valves or ball valves are another alternative. These are commonly found as the service shut-off valves in more-expensive water systems and usually found in gas taps. They can be identified by their range of motion—only 90°—between fully on and fully off. Usually, when the handle is in line with the pipe the valve is on, and when the handle is across the pipe it is closed. A cone valve consists of a shallowly-tapering cone in a tight-fitting socket placed across the flow of the fluid. In UK English this is usually known as a *taper-plug cock*. A ball valve uses a spherical ball instead. In either case, a hole through the cone or ball allows the fluid to pass if it is lined up with the openings in the socket through which the fluid enters and leaves; turning the cone using the handle rotates the passage away, presenting the fluid with the unbroken surface of the cone through which it cannot pass. Valves of this type using a cylinder rather than a cone are sometimes encountered, but using a cone allows a tight fit to be made even with moderate manufacturing tolerances. The ball in ball valves rotates within plastic seats.

Hands free infrared proximity sensors are replacing the standard valve. Thermostatically controlled electronic dual-purpose mixing or diverting valves are used within industrial applications to automatically provide liquids as required.

Foot controlled valves are installed within laboratory and healthcare/hospitals.

Modern taps often have aerators at the tip to help save water and reduce splashes. Without an aerator, water usually flows out of the tap in one big stream. An aerator spreads the water flow into many small droplets.

Modern bathroom and kitchen taps often use ceramic or plastic surfaces sliding against other spring-loaded ceramic surfaces or plastic washers. These tend to require far less maintenance than traditional globe valves and when maintenance is required, the entire interior of the valve is usually replaced, often as a single pre-assembled *cartridge*.

Of the trio of well-respected faucet manufacturers in North American plumbing circles, Moen and American Standard use cartridges (Moen's being O-ring based, American Standard's being ceramic), while Delta uses easily-replaced rubber seats facing the cartridge(s). Each design has its advantages: Moen cartridges tend to be easiest to find, American Standard cartridges have nearly infinite lifespan in sediment-free municipal water, and Delta's rubber seats tend to be most forgiving of sediment in well water.

Backflow: Most U.S. jurisdictions now require bibcocks to have a vacuum breaker or backflow preventer, so that water cannot return through the bibcock from the hose. This prevents contamination of the building or public water system should there be a pressure drop. In the UK, a double check valve is required to conform with water regulations; this is often incorporated within the body of the tap itself.

Tap Water

Tap water (*running water, city water, municipal water*, etc.) is a principal component of "indoor plumbing", which became available in urban areas of the developed world during the last quarter of the 19th century, and common during the mid-20th century. The application of technologies involved in providing clean or "potable" water to homes, businesses and public buildings is a major subfield of sanitary engineering.

The availability of tap water has major public health benefits, since it typically vastly reduces the risk to the public of contracting water-borne diseases. Providing tap water to large urban or suburban populations requires a complex and carefully designed system of collection, storage, treatment and distribution, and is commonly the responsibility of a government agency, often the same agency responsible for the removal and treatment of wastewater.

Specific chemical compounds are often added to tap water during the treatment process to adjust the pH or remove contaminants, as well as chlorine to kill biological toxins. Local geological conditions affecting groundwater are determining factors for the presence of various metal ions, often rendering the water "soft" or "hard".

Tap water remains susceptible to biological or chemical contamination. In the event of contamination deemed dangerous to public health, government officials typically issue an advisory regarding water consumption. In the case of biological contamination, residents are usually advised to boil their water before consumption or to use bottled water as an alternative. In the case of chemical contamination, residents may be advised to refrain from consuming tap water entirely until the matter is resolved.

In many areas a compound of fluoride is added to tap water in an effort to improve dental health among the public. In some communities "fluoridation" remains a controversial issue.

Potable Water Supply

This supply may come from several possible sources.

- Municipal water supply
- Water wells
- Delivered by truck
- Processed water from creeks, streams, rivers, lakes, rainwater, etc.

Domestic water systems have been evolving since people first located their homes near a running water supply, e.g., a stream or river. The water flow also allowed sending waste water away from the domiciles. Modern indoor plumbing delivers clean, safe, potable water to each service point in the distribution system. It is imperative that the clean water not be contaminated by the waste water (disposal) side of the process system. Historically, this contamination of drinking water has been the largest killer of humans.

Hot Water Supply

Domestic hot water is provided by means of water heater appliances, or through district heating. The hot water from these units is then piped to the various fixtures and appliances that require hot water, such as lavatories, sinks, bathtubs, showers, washing machines, and dishwashers.

Fixtures and Appliances

Everything in a building that uses water falls under one of two categories; Fixture or Appliance. As the consumption points above perform their function, most produce waste/sewage components that will require removal by the waste/sewage side of the system. The minimum is an air gap. See cross connection control & backflow prevention for an overview of backflow prevention methods and devices currently in use, both through the use of mechanical and physical principles. Fixtures are devices that use water without an additional source of power.

Pipe Materials

The earliest known evidence of drain tile being used for plumbing was found in Mesopotamia and is estimated to have been made around

3000 BC. The tiles were made from clay mixed with short lengths of straw. Both brass and copper pipes have been found in Egypt believed to have been made close to 2500 BC. The Romans made extensive use of lead pipe by joining sheets of lead into piping to carry their water supply and waste. During the Dark Ages following the fall of the Roman Empire, plumbing development virtually ceased for centuries except for isolated cases of plumbing installed in palaces and castles. In the 13th century, blacksmiths formed sheets of iron and lap welded the seam to create iron pipe. Though it is unclear as to when galvanized iron pipe was first used, a French chemist named Melouin is credited with developing the process in 1742. The earliest known use for cast iron pipe is for the water supply to a fountain in Langensalza, Germany, built around 1560. In 1819 the first cast iron pipe constructed in the US, was manufactured in Weymouth, New Jersey. Before that time, cast iron pipe and fittings had to be imported from Europe. It was not until the 1960s that the hubless cast iron pipe was brought to the U.S. from Europe by way of Canada. During the early 1900s, heavy-walled copper joined with threaded fittings was in use, but limited to public buildings because of its' high cost. However, during the 1930s light-gauge Copper tube and fittings were developed which made copper economically feasible and increased it's popularity. Polyvinyl Chloride(PVC) was produced experimentally in the 19th century but did not become practical to manufacture until 1926, when Waldo Semon of BF Goodrich Co. developed a method to plasticize PVC, making it easier to process. PVC pipe began to be manufactured in the 1940s and was in wide use during the DWV reconstruction of Germany and Japan following WWII. In the 1950s, plastics manufacturers in Western Europe and Japan began producing acrylonitrile butadiene styrene (ABS) pipe. The methods for producing cross-linked polyethylene (PEX) was also developed in the 1950s. Plastic supply pipes have become increasingly common, with a variety of materials and fittings employed, however plastic water pipes do not keep water as clean as copper and brass piping does. Copper pipe plumbing is bacteriostatic. This means that bacteria can't grow in the copper pipes. Plumbing codes define which materials may be used, and all materials must be proven by ASTM, UL, and/or NFPA testing.

Steel

Galvanized steel potable water supply and distribution pipes are commonly found with nominal diameters from 3/8" to 2". It is rarely used today for new construction residential plumbing. Steel pipe has

National Pipe Thread (NPT) standard tapered male threads, which connect with female tapered threads on elbows, tees, couplers, valves, and other fittings. Galvanized steel (often known simply as "galv" or "iron" in the plumbing trade) is relatively expensive, difficult to work with due to weight and requirement of a pipe threader. It remains in common use for repair of existing "galv" systems and to satisfy building code non-combustibility requirements typically found in hotels, apartment buildings and other commercial applications. It is also extremely durable.

Black lacquered steel pipe is the most widely used pipe material for fire sprinklers and natural gas. Most single family homes' systems typically won't require supply piping larger than 3/4". In addition to expense, another downside is it suffers from a tendency to become obstructed due to internal rusting and mineral deposits forming on the inside of the pipe over time after the internal galvanizing zinc coating has degraded. In potable water distribution service, galvanized steel pipe has a service life of about 30 to 50 years, although it is not uncommon for it to be less in geographic areas with corrosive water contaminants.

Lead Leaching

Generally, copper tubes are soldered directly into copper or brass fittings, although compression, crimp, or flare fittings are also used. Formerly, concerns with copper supply tubes included the lead used in the solder at joints (50% tin and 50% lead). Some studies have shown significant "leaching" of the lead into the potable water stream, particularly after long periods of low usage, followed by peak demand periods. In hard water applications, shortly after installation, the interior of the pipes will be coated with the deposited minerals that had been dissolved in the water, and therefore the vast majority of exposed lead is prevented from entering the potable water. Building codes now require lead-free solder. Building Codes throughout the U.S. require the use of virtually "lead-free" (<.2% lead) solder or filler metals in plumbing fittings and appliances as well.

Corrosion

Copper water tubes are susceptible to: cold water pitting caused by contamination of the pipe interior typically with soldering flux; erosion corrosion caused by high speed or turbulent flow; and stray current corrosion, caused by poor electrical wiring technique, such as improper grounding and bonding.

Pin Holes Due to Poor Plumbing Electrical Grounding And/or Bonding

Pin-hole leaks can occur anytime copper piping is improperly grounded and/or bonded; nonmetal piping, such as Pex or PVC, does not suffer from this problem. The phenomenon is known technically as *stray current corrosion* or *electrolytic pitting*. Pin-holing due to poor grounding or poor bonding occurs typically in homes where the original plumbing has been modified; homeowners may find a new plastic water filtration device or plastic repair union has interrupted the water pipe's electrical continuity to ground when they start seeing pinhole water leaks after a recent install. Damage occurs rapidly, usually being seen about six months after the ground interruption. Correctly installed plumbing appliances will have a copper bonding jumper cable connecting the interrupted pipe sections. Pinhole leaks from stray current corrosion can result in thousands of dollars in plumbing bills, and sometimes necessitating the replacement of the entire affected line. The cause is an electrical problem, not a plumbing problem; once the plumbing damage is repaired, an electrician should be consulted to evaluate the grounding and bonding of the entire plumbing system.

Stray current corrosion occurs because: 1) the piping system is connected accidentally or intentionally to a DC voltage source; 2) the piping does not have metal-to-metal electrical continuity; 3) if the voltage source is AC, one or more naturally occurring minerals coating the pipe interior act as a rectifier, converting AC current to DC. The DC voltage forces the water within the piping to act as an electrical conductor (an electrolyte). Electric current leaves the copper pipe, moves though the water across the nonconductive section, and reenters the pipe on the opposite side. Pitting occurs at the electrically negative side (the cathode), which may be upstream or downstream with respect to the water flow direction. Pitting occurs because the electrical voltage ionizes the pipe's interior copper metal, which reacts chemically with dissolved minerals in the water creating copper salts; these copper salts are soluble in water and wash away. Pits eventually grow and consolidate to form pin holes. Where there is one, there are almost certainly more.

Detecting and eliminating poor bonding is relatively straightforward. Detection is accomplished by use of a simple voltmeter set to DC with the leads placed in various places in the plumbing. Typically, a probe on a hot pipe and a probe on a cold pipe will tell you if there is improper grounding. Anything beyond a few millivolts is important, potentials of 200 mV are common. A missing bond will show up best in the area of the gap, as potential disperses as the water runs. Since the missing bond is usually seen near the water source, as filtration and treatment equipment are added, pinhole leaks can occur anywhere downstream. It is usually the cold water pipe, as this is the one that gets the treatment devices.

Correcting the problem is a simple matter of either purchasing a copper bonding jumper kit, composed of copper cable at least #6 AWG in diameter and two bronze ground clamps for affixing it the plumbing. See NFPA 70, the U.S. National Electrical Code Handbook (NEC), section on bonding and ground for details on selecting the correct bonding conductor wire size.

A similar bonding jumper wire can also be seen crossing gas meters, but for a different reason.

Note, if homeowners are experiencing shocks or sparks from plumbing fixtures or pipes, it is more than a missing bond, it is likely a live electrical wire is bridging to the plumbing and the plumbing system is not grounded. This is an electrical shock hazard and potential fire danger; consult an electrician immediately!

Plastics

Plastic pipe is in wide use for domestic water supply and drainage, waste, and vent (DWV) pipe. For example, polyvinyl chloride (PVC), chlorinated polyvinyl chloride (CPVC), polypropylene (PP), polybutylene (PB), and polyethylene (PE) may be allowed by code for certain uses. Some examples of plastics in water supply systems are:

- PVC/CPVC-rigid plastic pipes similar to PVC drain pipes but with thicker walls to deal with municipal water pressure, introduced around 1970. PVC should be used for cold water only, or venting. CPVC can be used for hot and cold potable water supply. Connections are made with primers and solvent cements as required by code.
- PP-The material is used primarily in housewares, food packaging, and clinical equipment, but since the early 1970s has seen increasing use worldwide for both domestic hot and cold water. PP pipes are heat fused, preventing the use of glues, solvents, or mechanical fittings. PP pipe is often used in green building projects.
- PBT-flexible (usually gray or black) plastic pipe which is attached to barbed fittings and secured in place with a copper crimp ring. The primary manufacturer of PBT tubing and fittings was driven into bankruptcy by a class-action lawsuit

over failures of this system. However, PB and PBT tubing has returned to the market and codes, typically first for 'exposed locations' such as risers.

- PEX-cross linked polyethylene system with mechanically joined fittings employing barbs and crimped steel or copper fittings.
- Polytanks-plastic polyethylene cisterns, underground water tanks, above ground water tanks, are made of linear polyethylene suitable as a potable water storage tank, provided in white, black or green, approved by NSF and made of FDA approved materials.
- Aqua-known as PEX-Al-PEX, for its PEX/aluminium sandwich-aluminium pipe sandwiched between layers of PEX and connected with brass compression fittings. In 2005, a large number of their fittings were recalled.

Fittings and Valves

Potable water supply systems require not only pipe, but also many fittings and valves which add considerably to their functionality as well as cost.

Regulation and Compliance

Before a water supply system is constructed or modified, the designer and contractor need to consult the local plumbing code and obtain a building permits prior to construction. Even replacing an existing water heater may require a permit and inspection of the work. NSF 61 is the U.S. national standard for potable water piping guidelines. National and local fire codes should be integrated in the design phase of the water system too to prevent "failure comply with regulations" notices. Some areas of the United States require on-site water reserves of potable and fire water by law.

Waste Water

The waste water from the various appliances, fixtures, and taps is transferred to the waste and sewage removal system via the sewage drain system. This system consists of larger diameter piping, water traps, and is well vented to prevent toxic gases from entering the living space.

Tap Water Versus Bottled Water

In modern Western society, levels of contaminants found in tap water vary for every household and plumbing system. A general conception regarding water is that bottled water is designed to be 'cleaner' than conventional tap water. However, in 1999, The Natural Resources Defense Council (NRDC) released controversial findings from a four year study on bottled water.

The results of this study claimed that one-third of the waters tested contained levels of contamination—including synthetic organic chemicals, bacteria, and arsenic—in at least one sample that exceeded allowable limits under either state or bottled water industry standards or guidelines.

However, the bottled water industry was quick to dispute the claim saying bottled water is one the most highly regulated food products under the FDA regulatory authority and that the FDA system worked extremely well when coupled with the International Bottled Water Association's Model Code and unannounced inspections.

Some municipalities in the United States are making an effort to use tap water over bottled water on government properties and events. However, others voted the idea down, including voters in the state of Washington who repealed a bottled water tax via citizen initiative.

James Workman, author of the book *Heart of Dryness: How the Last Bushmen Can Help Us Endure the Coming Age of Permanent Drought* and co-founder of SmartMarkets says that he doesn't believe that "tap water is bad and bottled water is good". Rather he cites differences in quality regulations and standards. "Bottled water is often tap water put through another filter and not held to the same quality regulations as public utility water is."

Chlorine is a disinfectant which is added to tap water in the United States. Chlorine can leave organic material like trihalomethanes and haloacetic acids in the water. The level of chlorine found is small, 1L of chlorinated water gives 0.2 mg of chlorine, which is too small to cause any health problems.

While most U.S. cities have what is considered safe tap water, contaminants ranging from bacteria to heavy metals are present in some tap water and violations of tap water standards have been well-publicized, such as the severe 1993 Cryptosporidium outbreak in Milwaukee, Wisconsin, which led to several deaths and around 400,000 illnesses. The University of Cincinnati recently completed a Tap Water Quality Analysis, funded by PUR, for major US cities.

Dissolved Gases

Tap water can sometimes appear cloudy, and this is often mistaken for a mineral impurity in the water. Cloudy water, also known as white water, is actually caused by air bubbles coming out of solution in the water. Because cold water holds more air than warm water, small bubbles will appear in water with a high dissolved oxygen content that is heated or depressurized, because this reduces how much dissolved gas the water can hold. This condition is completely harmless, and the cloudiness of the water disappears quickly as the gas is released from the water.

Water Pipe

Water pipes are pipes or tubes, frequently made of polyvinyl chloride (PVC/uPVC), ductile iron, steel, cast iron, polypropylene, polyethylene, or copper, that carry pressurized and treated fresh water to buildings (as part of a municipal water system), as well as inside the building.

Figure: *A system of copper water tubes used in a radiator heating system.*

For many centuries, lead was the favoured material for water pipes, due to its malleability (this use was so common that the word "plumbing" derives from the Latin word for lead). This was a source of lead-related health problems in the years before the health hazards of ingesting lead were fully understood; among these were stillbirths and high rates of infant mortality. Lead water pipes were still in common use in the early 20th century and remain in many households. Lead-tin alloy solder was commonly used to join copper pipes, but modern practice uses tin-antimony alloy solder to join copper in order to eliminate lead hazards.

Despite the Romans common use of lead pipes, they were rarely poisoned by them. Unlike other parts of the world where lead pipes cause poisoning, the Roman water had so much calcium in it, that a layer of plaque prevented the water contacting the lead itself. What often causes confusion is the large amount of evidence of widespread lead poisoning, particularly amongst those who would have had easy access to piped water. This was an unfortunate result of lead being used in cookware and as an additive to processed food and drink, such as a preservative in wine. Roman lead pipe inscriptions provided information on the owner to prevent water theft.

Figure: *Old water pipe, remnant of the Machine de Marly near Versailles, France*

Wooden pipes were used in London and elsewhere during the 16th and 17th centuries. The pipes were hollowed-out logs, which were tapered at the end with a small hole in which the water would pass through. The multiple pipes were then sealed together with hot animal fat. They were often used in Montreal and Boston in the 1800s, and built-up wooden tubes were widely used in the USA during the 20th century. These pipes, used in place of corrugated iron or reinforced concrete pipes, were made of sections cut from short lengths of wood. Locking of adjacent rings with hardwood dowel pins produced a flexible structure. About 100,000 feet of these wooden pipes were installed during WW2 in drainage culverts, storm sewers and conduits, under highways and at army camps, naval stations, airfields and ordnance plants.

Cast iron and ductile iron pipe was long a lower-cost alternative to copper, before the advent of durable plastic materials but special non-conductive fittings must be used where transitions are to be made to other metallic pipes, except for terminal fittings, in order to avoid corrosion owing to electrochemical reactions between dissimilar metals.

Bronze fittings and short pipe segments are commonly used in combination with various materials.

Pipe Us. Tube

The difference between pipe and tube is simply in the way it is sized. PVC pipe for plumbing applications and galvanized steel pipe for instance, are measured in IPS (iron pipe size). Copper tube, CPVC, PeX and other tubing is measured nominally, which is basically an average diameter. These sizing schemes allow for universal adaptation of transitional fittings. For instance, 1/2" PeX tubing is the same size as 1/2" copper tubing. 1/2" PVC on the other hand is not the same size as 1/2" tubing, and therefore requires either a threaded male or female adapter to connect them. When used in agricultural irrigation, the singular form "pipe" is often used as a plural.

Figure: *Typical PVC municipal water main being installed in Ontario, Canada*

Piping is available in rigid "joints", which come in various lengths depending on the material. Tubing, in particular copper, comes in rigid hard tempered "joints" or soft tempered (annealed) rolls. PeX and CPVC tubing also comes in rigid "joints" or flexible rolls. The temper of the copper, that is whether it is a rigid "joint" or flexible roll, does not affect the sizing.

The thicknesses of the water pipe and tube walls can vary. Pipe wall thickness is denoted by various schedules or for large bore polyethylene pipe in the UK by the Standard Dimension Ratio (SDR) being the ratio of the pipe diameter to its wall thickness. Pipe wall thickness increases with schedule, and is available in schedules 20, 40, 80, and higher in special cases. The schedule is largely determined by the operating pressure of the system, with higher pressures commanding greater thickness. Copper tubing is available in four wall thicknesses: type DWV (thinnest wall; only allowed as drain pipe per UPC), type 'M' (thin; typically only allowed as drain pipe by IPC code), type 'L' (thicker, standard duty for water lines and water service), and type 'K' (thickest, typically used underground between the main and the meter). Because piping and tubing are commodities, having a greater wall thickness implies higher initial cost. Thicker walled pipe generally implies greater durability and higher pressure tolerances.

Wall thickness does not affect pipe or tubing size. 1/2" L copper has the same outer diameter as 1/2" K or M copper. The same applies to pipe schedules. As a result, a slight increase in pressure losses is realized due to a decrease in flowpath as wall thickness is increased. In other words, 1 foot of 1/2" L copper has slightly less volume than 1 foot of 1/2 M copper.

Demand for copper products have fallen due to the dramatic increase in the price of copper, resulting in increased demand for alternative products including PEX and stainless steel.

Reinforced Thermoplastic Pipes

RTP is a generic term referring to a reelable high strength synthetic fibre (such as Glass, Aramid or Carbon) reinforced thermoplastic pipe, initially developed in the early 1990s by Wavin Repox, Akzo Nobel and by Tubes d'Aquitaine from France, who developed the first pipes reinforced with synthetic fibre to replace medium pressure steel pipes in response to growing demand for non-corrisive conduits for application in the onshore oil and gas industry, particularly in the Middle East.

Typically, the materials used in the construction of the pipe might be Polyethylene (PE), Polyamide-11 or PVDF and may be reinforced with Aramid or Polyester fibre although other combinations are used.

More recently the technology of producing such pipe, including the marketing, rests with a few key companies, one of which is Pipelife with Soluforce where it is available in coils up to 400 m (1,312 ft) length. These pipes are available in pressure ratings from 30 to 90 bar (3 to 9 MPa; 435 to 1,305 psi). Over the last few years this type of pipe has been acknowledged as a standard alternative solution to steel for oilfield flowline applications by certain oil companies and operators. The great advantage of this pipe is also its very fast installation time compared to steel pipe when considering the welding time as average speeds up to 1,000 m (3,281 ft)/day have been reached installing RTP in ground surface.

Primarily, the pipe provides benefit to applications where steel may rupture due to corrosion and installation time is an issue.

Technology and History

The idea of synthetic fibre reinforced pipe has origins in the flexible hose and offshore industry where it has been frequently used for applications such as control lines in umbilicals and production flowlines for over 30 years. However, the commercialisation and realisation of a competitive product for the onshore oil industry came from a partnership between Akzo Nobel (supplier of Aramid fibre) and Wavin Repox (manufacturer of reinforced thermoset pipes), where Bert Dalmolen initiated a project to develop such a pipe. He was later employed by Pipelife where a state of the art production line was developed to produce RTP. Pipelife also developed a pipe reinforced with steel wire to achieve even higher pressure ratings of over 150 bar (15 MPa; 2,176 psi) using steel reinforcement. Mr Chevrier (Tubes d'Aquitaine) also developed machinery that could produce such pipes, but was not successful in commercialising RTP.

Trap (Plumbing)

In plumbing, a trap is a U-, S-, or J-shaped pipe located below or within a plumbing fixture. An S-shaped trap is also known as the S-bend invented by Alexander Cummings in 1775 but became known as the U-bend following the introduction of the U-shaped trap by Thomas Crapper in 1880. The new U-bend could not jam, so, unlike the S-bend, it did not need an overflow. The bend is used to prevent sewer gases from entering buildings. In refinery applications, it also prevents hydrocarbons and other dangerous gases from escaping outside through drains.

The most common of these traps in houses is referred to as a P-trap. It is the addition of a 90 degree fitting on the outlet side of a U-bend, thereby creating a P-like shape. It can also be referred to as a sink trap due to the fact it is installed under most house sinks.

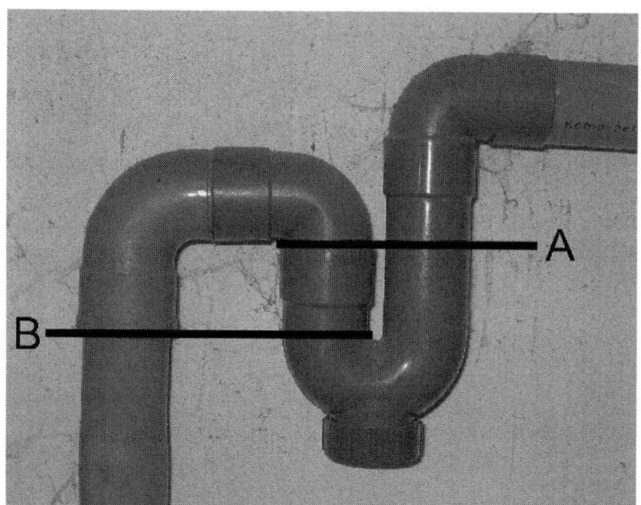

Figure: *Water seal in drain pipe under a sink. Water enters at right, fills the trap, and continues left. Siphoning takes place above the line "B", while inverted siphoning occurs below the line "A".*

Because of its shape, the trap retains a small amount of water after the fixture's use. This water in the trap creates a seal that prevents sewer gas from passing from the drain pipes back into the occupied space of the building. Essentially all plumbing fixtures including sinks, bathtubs, and toilets must be equipped with either an internal or external trap.

Because it is a localized low-point in the plumbing, sink traps also tend to capture heavy objects (such as jewelry) that are inadvertently dropped into the sink. Traps also tend to collect hair, sand, and other debris and limit the ultimate size of objects that will pass on into the rest of the plumbing, thereby catching over-sized objects. For all of these reasons, most traps can either be disassembled for cleaning or they provide some sort of cleanout feature.

Venting and Auxiliary Devices

Maintaining the water seal is critical to trap operation; traps can and do dry out, and poor venting can siphon or blow water out of the traps. This is usually avoided by venting the drain pipes downstream of the trap; by being vented to the atmosphere outside the building, the drain lines never operate at a pressure much higher or lower than atmospheric pressure. Plumbing codes usually provide strict limitations on how far a trap may be located from the nearest *vent stack.*

When a vent cannot be provided, codes may allow the use of an air admittance valve instead. These devices avoid negative pressure in the drain pipe by venting room air into the drain pipe (behind the trap). A "Chicago Loop" is another alternative.

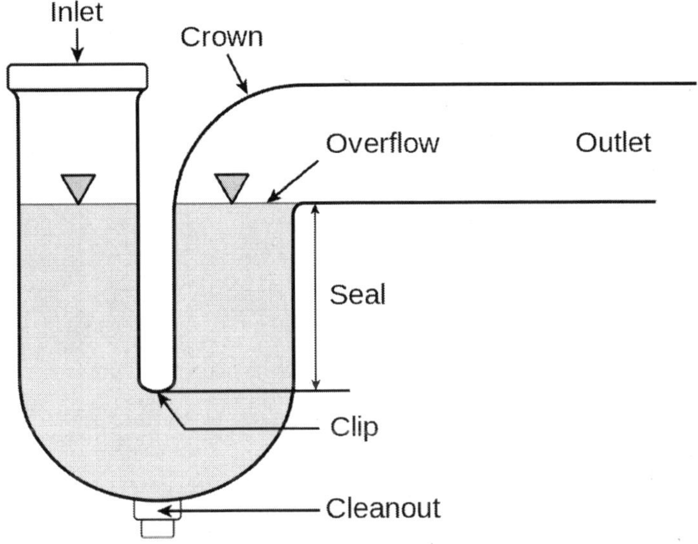

Figure: *Typical P-trap*

When a trap is installed on a fixture that is not routinely used, the eventual evaporation of the water in the trap must be considered. In these cases, a trap primer may be installed; these are devices that automatically recharge traps with water to maintain their water seals.

Accepted Traps

In many locations, "S" traps are no longer accepted by the plumbing codes and are even illegal, as these traps tend to easily siphon dry even when well-vented. It is often possible to tell whether a household uses an S or U-bend by checking for the presence of an overflow pipe outlet. Certain drum-styled traps are also discouraged or banned.

Buchan Trap

A Buchan Trap is a device made from fireclay located in a domestic sewer pipe to prevent vermin entering the pipe. The trap uses a water seal to prevent rats and mice climbing any further along the pipe. Waste flows from the house through a U-bend in the trap. This means that there is always water in the pipe preventing the passage of anything from the other direction. The device is a large clay U-bend with air-inlets and vents at the top. It is located below the ground level, but can be accessed through the air-inlet and a rodding hole. This hole allows drain rods to unblock anything located at the bottom of the U-bend.

Blockage Risk

The Buchan trap will collect solids, sludge and waste that is not in suspension and, over time, can clog up. The overflow will back up the pipe and exit from the lowest connected appliance (sink, bath, dish washer, washing machine, etc.). This can be a significant problem in multi-level dwellings (e.g. tenements) where those at the lowest level will receive all the waste waters from those above. They should be checked, and cleaned if required, on a regular basis (say annually). Traps of this type are widely believed by plumbers to stop the entry of rats but rats are good swimmers and can easiler pass through the short flooded section of the trap. The purpose of these traps (as this article says) was to help prevent sewer air entering houses-foul air was thought by the Victorians to be the source of disease.

The Buchan trap was devised in the 19th century to stop the flow of sewer gases also known as miasmas. It was believed that the disease cholera was an air born infection, not water born. The Buchan trap is normally found in the bottom of manholes or drop-shafts. It normally denotes the end point of the domestic property's sewer before it joins the main public sewer.

Drainage

Drainage is the natural or artificial removal of surface and sub-surface water from an area. Many agricultural soils need drainage to improve production or to manage water supplies.

Figure: *Remains of a drain at Lothal dating back to 2900BC*

The ancient Indus systems of sewerage and drainage that were developed and used in cities throughout the civilization were far more advanced than any found in contemporary urban sites in the Middle East and even more efficient than those in some areas of modern Pakistan and India today. All houses in the major cities of Harappa and Mohenjo-daro had access to water and drainage facilities. Waste water was directed to covered drains, which lined the major streets.

Drainage in the 19th Century

This operation is always best performed in spring or summer, when the ground is dry. Main drains ought to be made in every part of the field where a cross-cut or open drain was formerly wanted; they ought to be cut four feet (1.2 m) deep, upon an average. This completely secures them from the possibility of being damaged by the treading of horses or cattle, and being so far below the small drains, clears the water finely out of them.

Figure: *Tank Stream, a historical drain in the City of Sydney*

In every situation, pipe-turfs for the main drains, if they can be had, are preferable. If good stiff clay, a single row of pipe-turf; if sandy, a double row. When pipe-turf cannot be got conveniently, a good wedge drain may answer well, when the subsoil is a strong, stiff clay; but if the subsoil be only moderately so, a thorn drain, with couples below, will do still better; and if the subsoil is very sandy, except pipes can be had, it is in vain to attempt under-draining the fiel d by any other method. It may be necessary to mention here that the size of the main drains ought to be regulated according to the length and declivity of the run, and t eihther it can be he quantity of water to be carried off by them. It is always safe, however, to have the main drains large, and plenty of them; for economy here seldom turns out well.

Having finished the main drains, proceed next to make a small drain in every furrow of the field if the ridges formerly have not been less than fifteen feet (4.6 m) wide. But if that should be the case, first level the ridges, and make the drains in the best direction, and at such a distance from each other as may be thought necessary. If the water rises well in the bottom of the drains, they ought to be cut three feet (1 m) deep, and in this ease would dry the field sufficiently well, although they were from twenty-five to thirty feet (8 to 10 m) asunder; but if the water does not draw well to the bottom of the drains, two feet (0.6 m) will be a sufficient deepness for the pipe-drain, and two and a half feet (1 m) for the wedge drain. In no case ought they to be shallower where the field has been previously leveled. In this instance, however, as the surface water is carried off chiefly by the water sinking immediately into the top of the drains, it will be necessary to have the drains much nearer each other—say from fifteen to twenty feet (4.6 to 6 m). If the ridges are more than fifteen feet (4.6 m) wide, however broad and irregular they may be, follow invariably the line of the old furrows, as the best direction for the drains; and, where they are high-gathered ridges, from twenty to twenty-four inches will be a sufficient depth for the pipe-drain, and from twenty-four to thirty inches for the wedge-drain. Particular care should be taken in connecting the small and main drains together, so that the water may have a gentle declivity, with free access into the main drains.

When the drains are finished, the ridges are cleaved down upon the drains by the plough; and where they had been very high formerly, a second clearing may be given; but it is better not to level the ridges too much, for by allowing them to retain a little of their former shape, the ground being lowest immediately where the drains are, the surface water collects upon the top of the drains; and, by shrinking into them, gets freely away. After the field is thus finished, run the new ridges across the small drains, making them about ten feet (3 m) broad, and continue afterwards to plough the field in the same manner as dry land.

It is evident from the above method of draining that the expense will vary very much, according to the quantity of main drains necessary for the field, the distance of the small drains from each other, and the distance the turf is to be carried.

The advantage resulting from under-draining, is very great, for besides a considerable saving annually of water furrowing, cross cutting, etc., the land can often be ploughed and sown to advantage, both in the spring and in the fall of the year, when otherwise it would be found quite impracticable; every species of drilled crops, such as beans, potatoes, turnips, etc., can be cultivated successfully; and every species, both of green and white crops, is less apt to fail in wet and untoward seasons.

Wherever a burst of water appears in any particular spot, the sure and certain way of getting quit of such an evil is to dig hollow drains to such a depth below the surface as is required by the fall or level that can be gained, and by the quantity of water expected to proceed from the burst or spring. Having ascertained the extent of water to be carried off, taken the necessary levels, and cleared a mouth or loading passage for the water, begin the drain at the extremity next to that leader, and go on with the work till the top of the spring is touched, which probably will accomplish the intended object. But if it should not be completely accomplished, run off from the main drain with such a number of branches as may be required to intercept the water, and in this way disappointment will hardly be experienced. Drains, to be substantially useful, should seldom be less than three feet (1 m) in depth, twenty or twenty four inches thereof to be close packed with stones or wood, according to circumstances. The former are the best materials, but in many places are not to be got in sufficient quantities; recourse therefore, must often be made to the latter, though not so effectual or durable.

It is of vast importance to fill up drains as fast as they are dug out; because, if left open for any length of time, the earth is not only apt to fall in but the sides get into a broken, irregular state, which cannot afterwards be completely rectified. A proper covering of straw or sod should be put upon the top of the materials, to keep the surface earth from mixing with them; and where wood is the material used for filling up, a double degree of attention is necessary, otherwise the proposed improvement may be effectually frustrated.

The pit method of draining is a very effectual one, if executed with judgment. When it is sufficiently ascertained where the bed of water is deposited, which can easily be done by boring with an auger, sink a pit into the place of a size which will allow a man freely to work within its bounds. Dig this pit of such a depth as to reach the bed of the water meant to be carried off; and when this depth is attained, which is easily discerned by the rising of the water, fill up the pit with great land-stones and carry off the water by a stout drain to some adjoining ditch or mouth, whence it may proceed to the nearest river.

Current Practices

Modern drainage systems incorporate geotextile filters that retain and prevent fine grains of soil from passing into and clogging the drain. Geotextiles are synthetic textile fabrics specially manufactured for civil and environmental engineering applications. Geotextiles are designed to retain fine soil particles while allowing water to pass through. In a typical drainage system they would be laid along a trench which would then be filled with coarse granular material: gravel, sea shells, stone or rock. The geotextile is then folded over the top of the stone and the trench is then covered by soil. Groundwater seeps through the geotextile and flows within the stone to an outfall. In high groundwater conditions a perforated plastic (PVC or PE) pipe is laid along the base of the drain to increase the volume of water transported in the drain.

Alternatively, prefabricated plastic drainage systems made of HDPE called SmartDitch, often incorporating geotextile, coco fiber or rag filters can be considered. The use of these materials has become increasingly more common due to their ease of use which eliminates the need for transporting and laying stone drainage aggregate which is invariably more expensive than a synthetic drain and concrete liners.

Over the past 30 years geotextile and PVC filters have become the most commonly used soil filter media. They are cheap to produce and easy to lay, with factory controlled properties that ensure long term filtration performance even in fine silty soil conditions.

21st Century Alternatives

Seattle's Public Utilities created a pilot προγραμμε called Street Edge Alternatives (SEA Streets) Project. The project focuses on designing a system "to provide drainage that more closely mimics the natural landscape prior to development than traditional piped systems". The streets are characterized by ditches along the side of the roadway, with plantings designed throughout the area. An emphasis on non curbed sidewalks allows water to flow more freely into the areas of permeable surface on the side of the streets. Because of the plantings the run off water from the urban area does not all directly go into the ground but can also be absorbed into the surrounding environment. According to the monitoring by Seattle Public Utilities, they report a 99 percent reduction of storm water leaving the drainage project

Drainage in Construction

The civil engineer or site engineer is responsible for drainage in construction projects. They set out from the plans all the roads, street gutters, drainage, culverts and sewers involved in construction operations. During the construction of the work on site he/she will set out all the necessary levels for each of the previously mentioned factors.

Site engineers work alongside architects and construction managers, supervisors, planners, quantity surveyors, the general workforce, as well as subcontractors. Typically, most jurisdictions have some body of drainage law to govern to what degree a landowner can alter the drainage from his parcel.

Green Drain

Enzymatic drain cleaners can be a safer alternative to chemical drain cleaners, and are easier on the environment. They use bacteria or enzymes, which naturally feed on organic waste, such as hair and food waste that often block drains. These tiny organisms, then digest the waste and to recreate beneficial bacteria and enzymes throughout the septic system. In fact, drain cleaners enzyme originally used to clean septic tanks and sewage. Enzymatic drain cleaners are better for the environment because they prevent hazardous chemicals that may leak into soil and water from spreading.

Reasons for Artificial Drainage

Wetland soils may need drainage to be used for agriculture. In the northern USA and Europe, glaciation created numerous small lakes which gradually filled with humus to make marshes. Some of these were drained using open ditches and trenches to make mucklands, which are primarily used for high value crops such as vegetables.

The largest project of this type in the world has been in process for centuries in the Netherlands. The area between Amsterdam, Haarlem and Leiden was, in prehistoric times swampland and small lakes. Turf cutting (Peat mining), subsidence and shoreline erosion gradually caused the formation of one large lake, the Haarlemmermeer, or lake of Haarlem. The invention of wind powered pumping engines in the 15th century permitted drainage of some of the marginal land, but the final drainage of the lake had to await the design of large, steam powered pumps and agreements between regional authorities. The elimination of the lake occurred between 1849 and 1852, creating thousands of km^2 of new land.

Coastal plains and river deltas may have seasonally or permanently high water tables and must have drainage improvements if they are to be used for agriculture. An example is the flatwoods citrus-growing region of Florida. After periods of high rainfall, drainage pumps are employed to prevent damage to the citrus groves from overly wet soils. Rice production requires complete control of water, as fields need to be flooded or drained at different stages of the crop cycle. The Netherlands has also led the way in this type of drainage, not only to drain lowland along the shore, but actually pushing back the sea until the original nation has been greatly enlarged.

In moist climates, soils may be adequate for cropping with the exception that they become waterlogged for brief periods each year, from snow melt or from heavy rains. Soils that are predominantly clay will pass water very slowly downward, meanwhile plant roots suffocate because the excessive water around the roots eliminates air movement through the soil.

Other soils may have an impervious layer of mineralized soil, called a hardpan or relatively impervious rock layers may underlie shallow soils. Drainage is especially important in tree fruit production. Soils that are otherwise excellent may be waterlogged for a week of the year, which is sufficient to kill fruit trees and cost the productivity of the land until replacements can be established. In each of these cases appropriate drainage carries off temporary flushes of water to prevent damage to annual or perennial crops.

Drier areas are often farmed by irrigation, and one would not consider drainage necessary. However, irrigation water always contains minerals and salts, which can be concentrated to toxic levels by evapotranspiration. Irrigated land may need periodic flushes with excessive irrigation water and drainage to control soil salinity.

Plumbing Drainage Venting

In modern plumbing, a drain-waste-vent (or DWV) is part of a system that removes sewage and greywater from a building and regulates air pressure in the waste-system pipes, facilitating flow. Waste is produced at fixtures such as toilets, sinks and showers, and exits the fixtures through a trap, a dipped section of pipe that always contains water. All fixtures must contain traps to prevent sewer gases from leaking into the house. Through traps, all fixtures are connected to waste lines, which in turn take the waste to a soil stack, or soil vent pipe. At the building drain system's lowest point, the drain-waste vent is attached, and rises (usually inside a wall) to and out of the roof. Waste is removed from the building through the building drain and taken to a sewage line, which leads to a septic system or a public sewer. Cesspits are generally prohibited in developed areas.

The venting system, or plumbing vents, consists of pipes leading from waste pipes to the outdoors, usually through the roof. Vents provide a means to release sewer gases outside instead of inside the house. Vents also admit oxygen to the waste system to allow aerobic sewage digestion. Vents provide a way to equalize the pressure on both sides of trap, thereby allowing the trap to hold water, which is needed to maintain effectiveness of the trap. Every fixture is required to have an internal or external trap; double trapping is prohibited by plumbing codes. With exceptions, every plumbing fixture must have an attached vent. The top of stacks must be vented too, via a stack vent, which is sometimes called a stink pipe.

DWV systems maintain neutral air pressure in the drains, allowing flow of water and sewage down drains and through waste pipes by gravity. As such, it is critical that a downward slope be maintained throughout. In relatively rare situations, a downward slope out of a building to the sewer cannot be created, and a special collection pit and grinding lift 'sewage ejector' pump are needed. By contrast, potable water supply systems operate under pressure to distribute water up through buildings.

Purpose

A sewer pipe is normally at neutral air pressure compared to the surrounding atmosphere. When a column of waste water flows through a pipe, it compresses air in the pipe, creating a positive pressure that must be released or it will push back on the waste stream and downstream traps' water seals. As the column of water passes, air must flow in behind the waste stream or negative pressure results. The extent of these pressure fluctuations is determined by the fluid volume of the waste discharge.

Excessive negative air pressure, behind a 'slug' of water that is draining, can siphon water from trap seals at plumbing fixtures. Generally, a toilet outlet has the shortest trap seal, making it most vulnerable to being emptied by induced siphonage. An empty trap can allow noxious sewer gasses to enter a building.

On the other hand, if the air pressure within the drain becomes suddenly higher than ambient, this positive transient could cause waste water to be pushed into the fixture, breaking the trap seal, with dire hygiene and health consequences if too forceful. Tall buildings of three or more stories are particularly susceptible to this problem. Vent stacks are put in parallel to waste stacks to allow proper venting in tall buildings.

Venting Mechanisms

To prevent the problems of high pressure in a drain system, sewer pipes will usually vent via one of two mechanisms.

Venting to Atmosphere

Most residential buildings' drainage systems in North America are vented directly through the buildings' roofs. The DWV pipe is typically ABS or PVC DWV-rated plastic pipe equipped with a flashing to prevent rainwater from entering the buildings. Older homes may use copper, iron, lead or clay pipes, in rough order of increasing antiquity.

Under many older building codes, a vent stack, a pipe leading to the main roof vent, is required to be within a five foot radius of the draining fixture (sink, toilet, shower stall, etc.). To allow only one vent stack, and thus one roof protrusion as permitted by local building code, sub-vents may be tied together and exit a common vent stack.

A blocked vent is a relatively common problem caused by anything from leaves, to dead squirrels, to ice dams in very cold weather. Symptoms range from bubbles in the toilet bowl when it is flushed, to slow drainage, and all the way to siphoned (empty) traps and sewer gases entering the building. When a fixture trap is venting properly, a "sucking" sound can often be heard as the fixture empties out.

Air Admittance Valve

Air admittance valves (AAVs or Durgo valves) are negative pressure-activated, one-way mechanical vents, used in a plumbing system to eliminate the need for conventional pipe venting and roof penetrations. A discharge of wastewater causes the AAV to open, releasing the vacuum and allowing air to enter plumbing vent pipe for proper drainage. Since AAVs will only function under negative pressure situations they are not suitable for all venting applications, such as venting a sump, where positive pressures are created when the sump fills. Using AAVs can significantly reduce the amount of venting materials needed in a plumbing system, increase plumbing labour efficiency, allow greater flexibility in the layout of plumbing fixtures, and reduce long-term roof maintenance problems associated with conventional vent stack roofing penetrations.

While some state and local building departments prohibit AAVs, the International Residential and International Plumbing Codes allow it to be used in place of a vent-through-the-roof. AAV's are certified to reliably open and close a minimum of 500,000 times, (approximately 30 years of use) with no emanation of sewer gas; and some manufacturers claim their units are tested for up to 1.5 million cycles, or at least 80 years of use. Air Admittance Valves have been effectively used in Europe for more than two decades. US manufacturers offer warranties that range from 20 years to lifetime.

Fittings

Drainage and venting systems require not only pipe, but also many specialized fittings which add considerably to their cost of construction. But fittings such as "clean-outs" enhance the maintainability of the systems.

Tube Beading

Tube beading is a metal forming process that forms a bead on the end of a tube. Tube beads can be used to help hold a hose on the end of a tube or to strengthen the end of the tube. There are two forming processes: *internal roll forming* and *ram forming.*

Internal Roll Forming

Internal roll forming is generally slower than ram forming but it holds tight tolerances. Serrated clamp jaws are used to hold the tube while an axial pressure is applied to the end of the tube to form the bead.

Ram Forming

Ram forming is quicker and usually preferred when speed of production is a concern. The automotive industry usually uses this process over internal roll-forming. A clamp is used to hold the tube, the tube is expanded to desired diameter by an expansion punch, and then another punch is used to reduce the tube back to pilot diameter. Multiple beads are possible using this process.

Types of Pipes

Carbon and Alloy Steel Pipes

Carbon steel pipe (A53) is supplied by continuous buttweld or the electric-resistance welding process and is the most widely used pipe in use today. Carbon steel pipe is easily welded, threaded, machined and formed by means of machine or special forming fixture. Stocked in standard Schedule 40, other schedules available upon request.

Analysis

Carbon (C)	*Manganese (Mn)Max*	*Silicon (Si)Min*	*Phosphorus (P)Max*	*Sulfur (S)Max*
0.3	1.2	0.1	0.05	0.045

Schedule 40 pipe conforms to ASTM A53-B, ASTM A53-F

Mechanical Properties

Tensile Strength (PSI)	*Yield Strength (PSI)*	*Elongation in 2"*
60,000	35,000	20

Applications

Carbon pipe is typically used for conveyance of water, air, gas and steam; sprinkler systems, fencing, pipe piling, slurry lines, hydraulic hoists, pipe type cable systems, truck racks, frames, railings, ornamental works, and a multitude of structural applications

Size Conversion Schedule 40

Table

Pipe Size	*O.D. (in)*	*Wall (in)*	*Pipe Size*	*O.D. (in)*	*Wall (in)*
1/4	.540	.088	2-1/2	2.875	.203
3/8	.675	.091	3	3.500	.216
1/2	.840	.109	3-1/2	4.000	.226
3/4	1.050	.113	4	4.500	.237
1	1.315	.133	5	5.563	.258
1-1/4	1.660	.140	6	6.625	.280
1-1/2	1.900	.145	8	8.625	.322
2	2.375	.154	10	10.750	.365

Asbestos Cement Pipes

Asbestos cement pipes are made of a mixture of asbestos paste and cement compressed by steel rollers to form a laminated material of great strength and desnsity. its carrying capacity remains substantially constant as when first laid, irrespective of the quality of water. It can be drilled and tapped for connecting but does not have the same strength or suitability for threading as iron and any leakage in the thread will become worse as time passes. However this difficulty can be over come by screwing the ferrules through malleable iron saddles fixed at the point of service conncetions as is the general practice. The pipes are not suitable for use in sulphate soils.

The available safety against bursting under pressure and against failure in longitudinal bending, though less than that for spun iron pipes, is nevertheless adequate and increases as the pipe ages. In most cases good bedding of the pipes and the use of flexible joints are of greater importance in preventing failure by bending, than the strength of pipe itself. Flexible joints are used at regular intervals to provide for repairing of pipes if necessary.

AC pipes are manufactured from classes 5 to 25 and nominal diameters of 80 mm to 600 mm with the test pressure 5 to 25 Kg/ cm2.AC pipe can meet the general requirements of water supply undertaking for rising main as well as distribution main. It is classified as class 5,10,15,20 and 25, which have test pressures 5,10,15,20 and 25 Kgs/cm2. respectively. Working pressure shall not be greater than 50 % of test pressure for pumping mains and 67% for gravity mains.

Laying and Jointing

The width of the trech should be uniform throughout the length and greater than the out side diameter of the pipe by 300mm on either side of the pipe. The depth of the trench is usually kept 1 meter on the top of the pipe. For heavy traffic a cover of atleast 1.25 meter is provided on the top of the pipe.

The AC pipes to be laid are stacked along the trenches on the side or opposite to the spoils. Each pipe should be examined for any defects such as cracks. chipped ends, crusting of the sides etc. The defective pipes should be removed forthwith from the site as other wise they are likely to be mixed up with the good pipes. Before use the inside of the pipes will have to be cleaned. The lighter pipe weighing less than 80 Kg can be lowered in the trench by hand.

If the sides of the trench slope too much ropes must be used. The pipes of medium weight upto 200 Kg are lowered by means of ropes looped around both the ends. One ends of the rope is fastened to a wooden or steel stack driven into the ground and the other end of the rope is held by men and is slowly released to lower the pipe into the trench. After their being lowered into the trench they are aligned for jointing. The bed of the trench should be uniform.

Pipe Joints: There are two types of joints for AC pipes.

Cast Iron Detachable Joints

This consists of two cast iron flanges, a cast iron central collar and two rubber rings along with a set of nuts and bolts, for a particular joint. For this joint of the AC pipes should have flush ends. For jointing a flange, a rubber ring and a collar are slipped to the first pipe in that order; a flange and a rubber ring being introduced from the jointing of the next pipe. Both the pipes are now aligned and the collar centralised and the joints of the flanges tightened with nuts and bolts.

A.C Coupling Joint

This consists of an A.C Coupling and three special rubber rings. The pipes for these joints have chamfered ends. These rubber rings are positioned in the grooves inside the coupling, then grease is applied on the chamferred end and the pipe and coupling is pushed with the help of a jack against the pipe. The mouth of the pipe is then placed in the mouth of the coupling end and then pushed so as to bring the two chamfered ends close to each other. Wherever necessary change over from cast iron pipe to AC pipes or vice versa should be done with the help of suitable adapters IS 6530-1972 may be followed for laying A.C pipes.

Pressure Testing: The procedure for the test as adopted generally is as follows:

a) At a time one section of the pipe line between two sluice valves is taken up for testing. The section usually taken is about 500 meters long.

b) One of the valves is closed and the water is admitted into the pipe through the other, manipulating air valves suitably. (If there are no sluice valves in between the section, the end of the section can be sealed temporarily with an end cap having an outlet which can serve as an air relief vent or for filling the line as may be required. The pipe line after it is filled, should be allowed to stand for 24 hours before pressure testing).

c) After filling this sluice valve is closed and the pipe section is isolated.

d) Pressure gaugages are fitted at suitable intervals on the crown into the holes meant for the purpose.

e) The pipe section is then connected to the delivery side of the pump through a small valve.

f) The pump is then worked till the pressure inside reaches the designed value which can be read from the pressure gauges fixed.

g) After the required pressure has been attained, the valves is closed and the pump disconnected.

h) The pipe is then kept under the desired pressure during inspection for any defect i.e. leakages at the joints etc.

Flexible Hose and Tube

A hose is a hollow tube designed to carry fluids from one location to another. Hoses are also sometimes called *pipes* (the word *pipe* usually refers to a rigid tube, whereas a hose is usually a flexible one), or more generally *tubing*. The shape of a hose is usually cylindrical (having a circular cross section). Hose design is based on a combination of application and performance. Common factors are Size, Pressure Rating, Weight, Length, Straight hose or Coilhose and Chemical Compatabiltiy.

Hoses are made from one or a combination of many different materials. Applications mostly use nylon, polyurethane, polyethylene, PVC, or synthetic or natural rubbers, based on the environment and pressure rating needed. In recent years, hoses can also be manufactured from special grades of polyethylene (LDPE and especially LLDPE). Other hose materials include PTFE (Teflon), stainless steel and other metals.

Reinforced Rubber Hose

To achieve a better pressure resistance hoses can be reinforced with fibers or steel cord. Commonly used reinforcement methods are braiding, spiralling, knitting and wrapping of fabric plies. The reinfocement increases the pressure resistance but also the stiffness. To obtain flexibility corrugations or bellows are used. Usually circumferential or helical reinforcement rings are applied to maintain these corrugated or bellowed structures under internal pressure.

Applications

Hoses can be used in water or other liquid environments or to convey air or other gases. Hoses are used to carry fluids through air or fluid environments, and they are typically used with clamps, spigots, flanges, and nozzles to control fluid flow.

Specific applications include the following:

- A garden hose is used to water plants in a garden or lawn, or to convey water to a sprinkler for the same purpose.
- A Tough Hose is used to water crops in agriculture for drip irrigation
- A fire hose is used by firefighters to convey water to the site of a fire.
- Air hoses are used in underwater diving to carry air from a surface compressor or from air tanks. Industrial uses for operating flexible machinery and worktable tooling such as pneumatic screw drivers, grinders, staplers, etc.

- Hoses have been used in air brake systems ever since the technology was invented by George Westinghouse in 1868. This includes:
    - o Railway air brake hoses used between locomotives and railroad cars, and
    - o Truck air brake hoses used between tractors and semi-trailers.
- In building services, metal or plastic hoses are used to move water around a building; whilst air ducts are used to move air around. They can also be used to take out vibration, and thermal or settlement movement.
- Automotive hoses are used in automobiles to move fluids around for use in cooling, lubrication, and/or hydraulics. Hoses are also used to convey pressure or vacuum signals to control circuits or gauges, as well as conveying vacuum to heating, cooling, brake, and/or locking systems.
- In chemistry and medicine, hoses (usually called *tubes*) are used to move liquid chemicals or gases around.
- A fuel hose carries fuel.

Cut-off Factor

Cut-off factor (AKA "Cut-off Length") is a factor used to calculate the length of a hose cut to achieve the desired overall length of hose plus fittings. It is commonly seen in hydraulic hose and fitting specifications. The cut-off factor is specific to a particular hose fitting.

The formula used in calculating the optimum overall length is:

$$Cut\ Hose\ Length = Hose\ Assembly\ Overall\ Length - C1 - C2.$$

In this formula, C1 represents the cut-off factor of the first hose end and C2 represents the cut-off factor of the second hose end.

Tube (fluid Conveyance)

A tube, or tubing, is a long hollow cylinder used to convey fluids (liquids or gases). The terms "pipe" and "tube" are almost interchangeable, although minor distinctions exist — generally, a tube has tighter engineering requirements than a pipe. Both pipe and tube imply a level of rigidity and permanence, whereas a hose is usually portable and flexible. A tube and pipe may be specified by standard pipe size designations, *e.g.*, nominal pipe size, or by nominal outside or inside diameter and/or wall thickness. The actual dimensions of pipe are usually not the nominal dimensions: A 1-inch pipe will not actually measure 1 inch in either outside or inside diameter, whereas many types of tubing are specified by actual inside diameter, outside diameter, or wall thickness.

Manufacture

Tube drawing is a metalworking process to size tube by shrinking a large diameter tube into a smaller one, by drawing the tube through a die. This process produces high quality tubing with precise dimensions, good surface finish, and the added strength of cold working. Because it is so versatile, tube drawing is suitable for both large and small scale production. There are five types of tube drawing: tube sinking, mandrel drawing, stationary mandrel, moving mandrel, and floating mandrel. A mandrel is used in many of the types to prevent buckling or wrinkling in the workpiece.

Processes

Tube Sinking

Tube sinking, also known as free tube drawing, reduces the diameter of the tube without a mandrel inside the tube. The inner diameter (ID) is determined by the inner and outer diameter of the stock tube, the outer diameter of the final product, the length of the die landing, the amount of back tension, and the friction between the tube and the die. This type of drawing operation is the most economical, especially on thick-walled tubes and tubes smaller than 12 mm (0.47 in) in diameter, but does not give the best surface finish. As the tube thickness increases the surface finish quality decreases. This process is often used for the tubing on low cost lawn furniture.

Rod Drawing

Rod drawing is the process that draws the tube with a mandrel inside the tube; the mandrel is drawn with the tube. The advantage to this process is that the mandrel defines the ID and the surface finish and has a quick setup time for short runs. The disadvantages are that lengths are limited by the length of the mandrel, usually no more than 100 feet (30 m), and that a second operation is required to remove the mandrel, called *reeling*. This type of process is usually used on heavy walled or small ID tubes. Common applications include super-high pressure tubing and hydraulic tubing (with the addition of a finishing tube sinking operation). This process is also use for precision manufacturing of trombone handslides.

Fixed Plug Drawing

Fixed plug drawing, also known as stationary mandrel drawing, uses a mandrel at the end of the die to shape the ID of the tube. This process is slow and the area reductions are limited, but it gives the best inner surface finish of any of the processes. This is the oldest tube drawing method.

Floating Plug Drawing

Floating plug drawing, also known as floating mandrel drawing, uses a mandrel that is not anchored whatsoever to shape the ID of the tube. The mandrel is held in by the friction forces between the mandrel and the tube. This axial force is given by friction and pressure. The greatest advantage of this is that it can be used on extremely long lengths, sometimes up to 1,000 feet (300 m). The disadvantage is it requires a precise design otherwise it will give inadequate results. This process is often used for oil-well tubing.

Tethered Plug Drawing

Tethered plug drawing, also known as semi-floating mandrel drawing, is a mix between floating plug drawing and fixed plug drawing. The mandrel is allowed to float, but it still anchored via a tether. This process gives similar results to the floating plug process, except that it is designed for straight tubes. It gives a better inner surface finish than rod drawing.

There are three classes of manufactured tubing: seamless, as-welded or electric resistant welded (ERW), and drawn-over-mandrel (DOM).

- Seamless tubing is produced via extrusion or rotary piercing.
- Drawn-over-mandrel tubing is made from cold-drawn electrical-resistance-welded tube that is drawn through a die and over a mandrel to create such characteristics as dependable weld integrity, dimensional accuracy, and an excellent surface finish.

Standards

There are many industry and government standards for pipe and tubing. Many standards exist for tube manufacture; some of the most common are as follows:

- ASTM A213 Standard Specification for Seamless Ferritic and Austenitic Alloy-Steel Boiler, Superheater, and Heat-Exchanger Tubes
- ASTM A269 Standard Specification for Seamless and Welded Austenitic Stainless Steel Tubing for General Service

- ASTM A270 Standard Specification for Seamless and Welded Austenitic Stainless Steel Sanitary Tubing
- ASTM A511 Standard Specification for Seamless Stainless Steel Mechanical Tubing
- ASTM A513 Standard Specification for Electric-Resistance-Welded Carbon and Alloy Steel Mechanical Tubing
- ASTM A554 Standard Specification for Welded Stainless Steel Mechanical Tubing
- British Standard 1387:1985 Specification for screwed and socketed steel tubes and tubulars and for plain end steel tubes suitable for welding or for screwing to BS 21 pipe threads.

ASTM material specifications generally cover a variety of grades or types that indicate a specific material composition. Some of the most commonly used are:

- TP 304
- TP 316
- MT 304
- MT 403
- MT 506.

In installations using hydrogen, copper and stainless steel tubing must be factory pre-cleaned (ASTM B 280) and/or certified as instrument grade. This is due to hydrogen's particular propensities: to explode in the presence of oxygen, oxygenation sources, or contaminants; to leak due to its atomic size; and to cause embrittlement of metals, particularly under pressure.

Calculation of Strength

For a tube of silicone rubber with a tensile strength of 10 MPa and a 8 mm outer diameter and 2 mm thick walls. The maximum pressure may be calculated as follows:

Outer diameter = 0.008 [meter]

Wall thickness = 0.002 [meter]

Tensile strength = 10 * 1000000 [Pa]

Pressure burst = (Tensile strength * Wall thickness * 2/(10 * Outer diameter)) * 10 [Pa]

Gives burst pressure of 5 MPa.

Using a safety factor:

Pressure max = (Tensile strength * Wall thickness * 2/(10 * Outer diameter)) * 10/Safety_factor [Pa].

Chapter 4

Gas Networks Simulation

Gas networks simulation is a process of defining the mathematical model of the gas transport and distribution systems, which are usually composed of highly integrated pipe networks operating over a wide range of pressures. Simulation allows to predict the behaviour of gas network systems under different conditions. Such predictions can be effectively used to guide decisions regarding the design and operation of the real system.

Simulation Types

Depending on the gas flow characteristics in the system there are two states that can be matter of simulation:

- Steady state-the simulation does not take into account the gas flow characteristics' variations over time and described by the system of algebraic equations, in general nonlinear ones.
- Unsteady state (transient flow analysis)-described either by a partial differential equation or a system of such equations. Gas flow characteristics are mainly functions of time.

Network Topology

In the gas networks simulation and analysis, matrices turned out to be the natural way of expressing the problem. Any network can be described by set of matrices based on the network topology. The network consists of one source node (reference node) L1, four load nodes (2, 3, 4 and 5) and seven pipes or branches. For network analysis it is necessary to select at least one reference node. Mathematically, the reference node is referred to as the independent node and all nodal and branch quantities are dependent on it. The pressure at source node is usually known, and this node is often used as the reference node. However, any node in the network may have its pressure defined and can be used as the reference node. A network may contain several sources or other pressure-defined nodes and these form a set of reference nodes for the network.

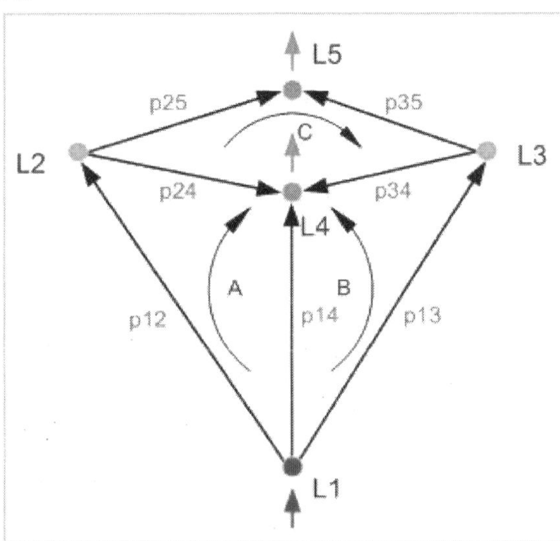

The load nodes are points in the network where load values are known. These loads may be positive, negative or zero. A negative load represents a demand for gas from the network. This may consist in supplying domestic or commercial consumers, filling gas storage holders, or even accounting for leakage in the network. A positive load represents a supply of gas to the network. This may consist in taking gas from storage, source or from another network. A zero load is placed on nodes that do not have a load but are used to represent a point of change in the network topology, such as the junction of several branches. For steady state conditions, the total load on the network is balanced by the inflow into the network at the source node.

The interconnection of a network can produce a closed path of branches, know as a loop. In figure, loop A consists of branches p12-p24-p14, loop B consists of p13-p34-p14, and loop C consists of p24-p25-p35-p34. A fourth loop may be defined as p12-p24-p34-p13, but it is redundant if loops A, B and C are also defined. Loops A, B and C are independent ones but the fourth one is not, as it can be derived from A, B and C by eliminating common branches.

To define the network topology completely it is necessary to assign a direction to each branch. Each branch direction is assigned arbitrarily and is assumed to be positive direction of flow in the branch. If the flow has the negative value, then the direction of flow is opposite to branch direction. In the similar way, direction is assigned to each loop and flow in the loop.

The solutions of problems involving gas network computation of any topology requires such a representation of the network to be found which enables the calculations to be performed in the most simple way. These requirements are met by the graph theory which permits representation of the network structure by means of the incidence properties of the network components and, in consequence, makes such a representation explicit.

Flow Equations

The calculation of the pressure drop along the individual pipes of a gas network requires use of the flow equations. Many gas flow equations have been developed and a number have been used by the gas industry. Most are based on the result of gas flow experiments. The result of the particular formula normally varies because these experiments were conducted over different range of flow conditions, and on varying internal surface roughness. Instead, each formula is applicable to a limited range of flow and pipe surface conditions.

Mathematical Methods of Simulation

Steady State Analysis

A gas network is in the steady state when the values of gas flow characteristics are independent of time and system described by the set of nonlinear equations. The goal of simple simulation of a gas network is usually that of computing the values of nodes' pressures, loads and the values of flows in the individual pipes. The pressures at the nodes and the flow rates in the pipes must satisfy the flow equations, and together with nodes' loads must fulfil the first and second Kirchhoff's laws. There are many methods of analysing the mathematical models of gas networks but they can be divided into two types as the networks, the solvers for low pressure networks and solvers for high pressure networks.

The networks equations are nonlinear and are generally solved by some of Newton iteration; rather than use the full set of variables it is possible to eliminate some of them. Based on the type of elimination we can get solution techniques are termed either nodal or loop methods.

Newton-Nodal Method

The method is based on the set of the nodal equations which are simply mathematical representation of Kirchhof's first law which states that the inlet and outlet flow at each node should be equal. Initial approximation is made to the nodal pressures. The approximation is then successively corrected until the final solution is reached.

Disadvantages

- Poor convergence characteristics, the method is extremely sensitive to initial conditions.

Advantages

- Does not require extra computation to produce and optimize a set of loops.
- Can easily adapted for optimization tasks.

Newton-loop Method

The method is based on the generated loops and the equations are simply mathematical representation of Kirchhof's second law which states that the sum of the pressure-drops around any loop should be zero. Before using loops method the fundamental set of loops need to be found. Basically the fundamental set of loops can be found by constructing spanning tree for the network. The standard methods for producing spanning tree is based on a breadth-first search or on a depth-first search which are not so efficient for large networks, because the computing time of these methods is proportional to n^2, where n is the number of pipes in the network. More efficient method for large networks is the forest method and its computational time is proportional to $n*\log_2 n$.

The loops that are produced from the spanning tree are not the best set that could be produced. There is often significant overlap between loops with some pipes shared between several loops. This usually slows convergence, therefore the loops' reduction algorithm needs to be applied to minimize the loops overlapping. This is usually performed by replacing the loops in the original fundamental set by smaller loops produced by linear combination of the original set.

Disadvantages

- It requires extra computation to produce and optimize a set of loops.
- The dimension of the equations to be solved is smaller but they are much less sparse.

Advantages

- The main advantage is that the equation can be solved very efficiently with an iterative method that avoids the need of matrix factorization and consequently has a minimal requirement for storage; this makes it very attractive for low pressure networks with a large number of pipes.

- Fast convergence which is less sensitive to the initial conditions.

Newton Loop-Node Method

The Newton loop-node method is based on Kirchhoff's first and second laws. The Newton loop-node method is the combination of the Newton nodal and loop methods and does not solve loop equations explicitly. The loop equations are transformed to an equivalent set of nodal equations, which are then solved to yield the nodal pressures. The nodal pressures are used then to calculate the corrections to the chord flows (which is synonymous to loop flows), and the tree branch flows are obtained from them.

Disadvantages

- Since set of nodal equations are solved nodal Jacobi matrix is used which is more sparse then the equivalent loop Jacobi matrix which may have negative impact on computational efficiency.

Advantages

- Good convergence characteristics of loop method are maintained.
- No need to define and optimize the loops.

Computer Simulation

The importance of the mathematical methods' efficiency arises from the large scale of simulated network. It is required that the computation costs of the simulation method be low, this is related to the computation time and computer storage. At the same time the accuracy of the computed values must acceptable for the particular model.

Operation Pluto

Operation *Pluto* (Pipe-Lines Under The Ocean) was a World War II operation by British scientists, oil companies and armed forces to construct undersea oil pipelines under the English Channel between England and France.

The scheme was developed by Arthur Hartley, chief engineer with the Anglo-Iranian Oil Company. Allied forces on the European continent required a tremendous amount of fuel. Pipelines were considered necessary to relieve dependence on oil tankers, which could be slowed by bad weather, were susceptible to German submarines, and were also needed in the Pacific War. Geoffrey William Lloyd, the Minister for Petroleum, met Admiral Mountbatten, Chief of Combined Operations, whose area this was, in 1942 and then the Chairman of Anglo-Iranian. Hartley's idea of using adapted submarine telephone cable was adopted.

Development

Two types of pipeline were developed: the flexible *HAIS* pipe with a 3 inch (75 mm) diameter lead core, weighing around 55 long tons per nautical mile (30 t/km), was essentially a development by Siemens Brothers, (in conjunction with the National Physical Laboratory) of their existing undersea telegraph cables, and known as HAIS from Hartley-Anglo-Iranian-Siemens.

The second type was a less flexible steel pipe of similar diameter, developed by engineers from the Iraq Petroleum Company and the Burmah Oil Company, known as *HAMEL* from the contraction of the two chief engineers, HA Hammick and BJ Ellis. It was discovered in testing that the HAMEL pipe was best used with final sections of HAIS pipe each end. Because of the rigidity of the HAMEL pipe, a special apparatus code-named *The ConunDrum* was developed (Picture) to lay the pipe.

The first prototypes were tested in May 1942 (across the River Medway), and in June in deep water across the Firth of Clyde, before going into production. Manufacturing was carried out by Siemens Brothers at Woolwich, Henley's at North Woolwich, Callender's at Erith and Standard Telephones and Cables at Greenwich. Because of capacity limitations in the UK, some HAIS pipeline was also manufactured in the United States.

In June 1942 the Post Office cable ship *Iris* laid lengths of both Siemens' and Henleys' cable in the Clyde. Both pipelines were completely successful and PLUTO was formally brought into the plans for the invasion of Europe. The project was deemed "strategically important, tactically adventurous, and, from the industrial point of view, strenuous".

The Clyde trials showed that it was necessary to maintain an internal pressure of about 7 bar (100 pounds/in^2) in the pipeline at all times, even during manufacture. Also, existing cable ships were not large enough, nor were their loading and laying gear sufficiently powerful and robust. Consequently a number of merchant ships were converted to pipe-laying by stripping the interiors and building in large cylindrical steel tanks, fitting special hauling gear and suitable sheaves and guides. It was to the specialised Johnson and Phillips company that the Petroleum Warfare Department turned for special gear to handle and lay the pipe. As the pipe could not be bent to a smaller radius than five feet; a new haul-off drum of ten-feet diameter and fleeting ring, together with roller type bow and stern gear, were produced, and the final equipment fitted to HMS *Holdfast*.

Full-scale production of the two-inch pipe was started on the 14th August 1942, using steel from the now near defunct Corby steel works, and six weeks later, on the 30th October a thirty-mile length was loaded on board HMS *Holdfast* under the command of Commander Treby-Heale OBE, RNR, which was to be used as a full-scale rehearsal of Operation PLUTO. This trial took place between 26 December and 30 December, the thirty-mile length being laid across the Bristol Channel, in very bad and rough weather, and the shore ends being connected up at Swansea and Ilfracombe. Those on board to monitor the test were Mr. Hartley (Anglo-Iranian Oil), Mr. Tombs (Anglo-Iranian Oil), Mr. Colby (Iraq Petroleum), Mr. Betson (Post Office), Commander Hardy (Admiralty) and Mr. Whitehead OBE (Johnson and Phillips), who had designed the pipe handling equipment.

The rehearsal was a success, so much so that a three-inch (76 mm) diameter pipe rather than two was considered. This reduced the number of pipelines needed to pump the planned volume of petrol across the channel. This decision necessitated further alterations and additions to the pipeline handling gear. Two further ships were equipped with handling gear, these being HMS *Sancroft* and HMS *Latimer*,(later renamed as *Empire Baffin* and *Empire Ridley* reespectively) both of which could handle 100 miles (160 km) of three-inch (76 mm) pipe weighing approximately 6,000 tons.

The pipeline across the Bristol Channel was used to supply parts of Devon and Cornwall for the next year, during which time RASC and RE army personnel were trained to petrol pumping equipment in readiness for the invasion of Europe.

Johnson and Phillips were asked to provide storage sites in the East India and Surrey Commercial Docks. These sites were obtained and equipped with tubular steel bridges with overhead hauling gear erected in such a position that the pipe could be taken from a ship's tanks.

Placement

After full-scale testing of a 83 km (45 nautical mile) HAIS pipe between Swansea in Wales and Watermouth in North Devon, the first line to France was laid on August 12, 1944, over the 130 km (70 nautical miles) from Shanklin Chine on the Isle of Wight through the English Channel to Cherbourg. A further HAIS pipe and two HAMELs followed. As the fighting moved closer to Germany, 17 other lines (11 HAIS and 6 HAMEL) were laid from Dungeness to Ambleteuse in the Pas-de-Calais.

Figure: *A surviving section of the pipeline at Shanklin Chine.*

The *PLUTO* Pipelines were linked to pump stations on the English coast, housed in various inconspicuous buildings including cottages and garages. Though uninhabited, these were intended to cloak the real purpose of the buildings. Pluto Cottage at Dungeness, a pumping station built to look like a small house, is now a Bed and Breakfast. In England, the *PLUTO* pipelines were supplied by a 1,609 km (1,000 mi) network of pipelines (constructed at night to prevent detection by aerial reconnaissance) to transport fuel from ports including Liverpool and Bristol. In Europe, the pipelines were extended as the troops moved forward and eventually reached as far as the Rhine.

In January 1945, 305 tonnes (300 long tons) of fuel was pumped to France per day, which increased tenfold to 3,048 tonnes (3,000 long tons) per day in March, and eventually to 4,000 tons (almost 1,000,000

Imperial gallons) per day. In total, over 781 000 m^3 (equal to a cube with 92 meter long sides or over 172 million imperial gallons) of gasoline had been pumped to the Allied forces in Europe by VE day, providing a critical supply of fuel until a more permanent arrangement was made, although the pipeline remained in operation for some time after.

Figure: *PLUTO Pump from Sandown on the Isle of Wight*

Dumbo was the codename given to the pipeline that ran across Romney Marsh to Dungeness and then across the English Channel to France. The route of the pipeline can be traced in various places on Romney Marsh. Where the pipeline crossed water drainage ditches it ran above ground in a concrete case. Several of these can still be found.

Along with the Mulberry Harbours that were constructed immediately after D-Day, Operation *Pluto* is considered one of history's greatest feats of military engineering. The pipelines are also the forerunners of all flexible pipes used in the development of offshore oil fields.

In Film

In 1994, the Midland Bank (now part of HSBC) sponsored a black-and-white film which contained a remarkable amount of historical archive film showing the entire history and construction of the *Pluto* Project, the HAIS pipe and the Conundrum reels. It mentions the codewords *Bambi*, *Watson*, *Dumbo Near*, *Dumbo Far* and other terminal names. It shows how the HAIS pipe was constructed and increased in diameter from about 2" to the later 3" operational size. When the landing site for the invasion was switched from Calais to Normandy, the pipeline needed to be increased from its original length to around 70 miles (110 km), and the film tells of how the American pipeline industry became involved in producing the extra amount of HAIS pipe.

The film can be seen in a small heritage museum at Shanklin Chine in the Isle of Wight, one of the *Pluto* terminals, where there are also a lot of other memorabilia, books and photographs. One of the original pumps used on the Isle of Wight is preserved in the Bembridge Heritage Centre. Brenzett Museum, Romney Marsh, houses a small permanent display related to Pluto at Dungeness.

A film entitled *Operation Pluto* produced by the former Ministry of Fuel and Power is held in the British Film Institute archives. This film was part of a loop of films that was shown at the East Carlton Park steel heritage centre in Northamptonshire for many years.

Pipeline Bridge

A pipeline bridge is a bridge for running a pipeline over a river. Pipeline bridges for liquids and gases are, as a rule, only built when it is not possible to run the pipeline on a conventional bridge or under the river. However, as it is more common to run pipelines for centralized heating systems overhead, for this application even small pipeline bridges are common.

Types

As, in pipelines, there is normally a steady flow, they can be designed as suspension bridges.

Walkways

A pipeline bridge may be equipped with a walkway for maintenance purposes but, in most cases, this is not open for public access for safety and security reasons.

Pipeline Integrity Management

The concept of pipeline integrity management became especially important as a result of the Pipeline Safety Improvement Act of 2002 signed by President Bush on December, 12 2002.

History

2001 Liquid Integrity Management Program (LIMP) came into law, 2003 Transmission Integrity Management Program (TIMP) came into law, 2008 The final rule is expected for Distribution Integrity Management Program (DIMP).

Pipeline Integrity Management is process of managing pipeline operational risk and mitigating it affectively to ensure pipeline assets are maintained in safe and reliable condition.

Pipeline failures generally result in high incident events that can have devastating impact on life, reputation, production and environment. The management of this threat is a critical component of any safe operating asset

Integrity Management Program

The Integrity Management Program (IMP) provides a means to improve the safety of pipeline systems and to allocate resources effectively to:

- Identify and analyse actual and potential precursor events that can result in pipeline transport incidents.
- Provide a comprehensive and integrated means for examining and comparing the spectrum of risks and risk reduction activities available.
- Provide a structured and comprehensive means for selecting and implementing risk reduction activities.
- Ensure the utilization of appropriately trained and qualified company and third-party personnel for risk management and pipeline integrity assurance.

Pipeline Risk Management Information System

Pipeline Risk Management Information System (PRIMIS) Integrity Management Programs have led to a reduced amount of pipeline accidents. These were originally created for transmission pipelines.

Pneumatic Tube, a Method for Sending Documents and Other Solid Materials in Capsules Through a Tube

Pneumatic tubes (or capsule pipelines; also known as Pneumatic Tube Transport or PTT) are systems in which cylindrical containers are propelled through a network of tubes by compressed air or by partial vacuum. They are used for transporting solid objects, as opposed to conventional pipelines, which transport fluids. Pneumatic tube networks gained great prominence in the late 19th and early 20th century for businesses or administrations that needed to transport small but urgent packages (such as mail or money) over relatively short distances (within a building, or, at most, within a city). Some of these systems grew to great complexity, but they were eventually superseded by more modern methods of communication and courier transport, and are now much rarer than before. However, in some settings, such as hospitals, they remain of great use, and have been extended and developed further technologically in recent decades.

A small number of pneumatic transportation systems were also built for larger cargo, to compete with more standard train and subway systems. However, these never gained as much popularity as practical systems.

Historical Use

Pneumatics can be traced back to Hero of Alexandria in the 1st century AD, though there was apparently no thought of using them to move objects through pipes.

Pneumatic capsule transportation was originally invented by William Murdoch. Though a marvel of the time, and a successful sideshow, it was considered little more than a novelty until the invention of the capsule in 1836. The Victorians were the first to use *capsule pipelines* to transmit telegraph messages, or telegrams, to nearby buildings from telegraph stations.

While they are commonly used for small parcels and documents — now most often used as cash carriers at banks or supermarkets — they were originally proposed in the early 19th century for transport of heavy freight. It was once envisioned that networks of these massive tubes might be used to transport people.

Current Use

The technology is still used on a smaller scale. While its use for communicating information has been completely superseded by electronic systems, pneumatic tubes are still widely used for transporting small valuable objects, or where convenience and speed in a local environment is useful.

In the United States, a large number of drive-up banks use pneumatic tubes to transport cash and documents between cars and tellers. Most hospitals have a computer-controlled pneumatic tube system to deliver drugs, documents and specimens to and from laboratories and nurses' stations. Many factories use them to deliver parts quickly across large campuses. Many larger stores use systems to securely transport excess cash from checkout stands to back offices, and to send change back to cashiers. NASA's original Mission Control Center in Houston, Texas had pneumatic tubes connecting controller consoles with staff support rooms. Denver International Airport is noteworthy for the large number of pneumatic tube systems, including a 25 cm diameter system for moving aircraft parts to remote concourses, a 10 cm system for United Airlines ticketing, and a robust system in the parking toll collection system with an outlet at every booth.

Pneumatic tube systems are used in science, to transport samples during neutron activation analysis. Samples must be moved from the nuclear reactor core, in which they are bombarded with neutrons, to the instrument that records the resulting radiation. As some of the radioactive isotopes in the sample can have very short half-lives, speed is important. These systems may be automated, with a magazine of sample tubes that are moved into the reactor core in turn for a predetermined time, before being moved to the instrument station and finally to a container for storage and disposal.

Until it closed in early 2011, a McDonald's in Edina, MN claimed on its receipts to be the "World's Only Pneumatic Air Drive-Thru," sending food from their strip-mall location to a drive-through in the middle of a parking lot.

In Britain, the House of Commons telephone and computer exchange also has a pneumatic tube system in place.

Usages

In Postal Service

Pneumatic post or pneumatic mail is a system to deliver letters through pressurized air tubes. It was invented by the Scottish engineer William Murdoch in the 19th century and was later developed by the London Pneumatic Despatch Company. Pneumatic post systems were used in several large cities starting in the second half of the 19th century (including an 1866 London system powerful and large enough to transport humans during trial runs-though not intended for the purpose), but were largely abandoned during the 20th century.

It was also speculated that a system of tubes might deliver mail to every home in the US. A major network of tubes in Paris was in use until 1984, when it was finally abandoned in favor of computers and fax machines. In Prague, in the Czech Republic, a network of tubes extending approximately 60 kilometres in length still exists for delivering mail and parcels. Following the 2002 European floods, the Prague system sustained damage, and operation was halted indefinitely.

Pneumatic post stations usually connected post offices, stock exchanges, banks and ministries. Italy was the only country to issue postage stamps (between 1913 and 1966) specifically for pneumatic post. Austria, France, and Germany issued postal stationery for pneumatic use.

Typical current applications are in banks and hospitals. Many large retailers use pneumatic tubes to transport cheques or other documents from cashiers to the accounting office.

Historical Use

- 1853: linking the London Stock Exchange to the city's main telegraph station (a distance of 220 yards (200 m))
- 1861: in London with the London Pneumatic Despatch Company providing services from Euston railway station to the General Post Office and Holborn
- 1865: in Berlin (until 1976), the *Rohrpost*, a system 400 kilometers in total length at its peak in 1940
- 1866: in Paris (until 1984, 467 kilometers in total length from 1934)
- 1875: in Vienna (until 1956)
- 1887: in Prague (until 2002 due to flooding), the Prague pneumatic post
- 1897: in New York City (until 1953).
- other cities: Munich, Rio de Janeiro, Buenos Aires, Hamburg, Rome, Naples, Milan, Marseilles, Melbourne, Boston, Philadelphia, Chicago, St. Louis

In Public Transportation

19th Century

In 1812, George Medhurst first proposed, but never implemented, blowing passenger carriages through a tunnel. Precursors of pneumatic tube systems for passenger transport, the atmospheric railway (for which the tube was laid between the rails, with a piston running in it suspended from the train through a sealable slot in the top of the tube) were operated as follows:

- 1844–54: Dublin and Kingstown Railway's Dalkey Atmospheric Railway between Kingstown (Dún Laoghaire) and Dalkey, Ireland (1.75 mi (2.82 km))
- 1846–47: London and Croydon Railway between Croydon and New Cross, London, England (7.5 mi (12.1 km))
- 1847–48: Isambard Kingdom Brunel's South Devon Railway between Exeter and Newton Abbot, England (20 mi (32 km))

- 1847–60: Paris–Saint-Germain railway between Bois de Vésinet and Saint-Germain-en-Laye, France (2 km (1.2 mi))

In 1861, the London Pneumatic Despatch Company built a system large enough to move a person, although it was intended for parcels. The inauguration of the new Holborn Station on 10 October 1865 was marked by having the Duke of Buckingham, the chairman, and some of the directors of the company blown through the tube to Euston (a five minute trip).

The 550-meter Crystal Palace pneumatic railway was exhibited at the Crystal Palace in 1864. This was a prototype for a proposed *Waterloo and Whitehall Railway* that would have run under the River Thames linking Waterloo and Charing Cross. Digging commenced in 1865 but was halted in 1868 due to financial problems.

In 1867 at the American Institute Fair in New York, Alfred Ely Beach demonstrated a 32.6 m long, 1.8 m diameter pipe that was capable of moving 12 passengers plus a conductor. In 1869, the Beach Pneumatic Transit Company of New York secretly constructed a 95 m long, 2.7 m diameter pneumatic subway line under Broadway, to demonstrate the possibilities of the new transport mode.

The line only operated for a few months, closing after Beach was unsuccessful in getting permission to extend it – Boss Tweed, an immensely powerful local politician, did not want it to go ahead as he was intending to personally invest into competing schemes for an elevated rail line.

20th Century

In the 1960s, Lockheed and MIT with the United States Department of Commerce conducted feasibility studies on a vactrain system powered by ambient atmospheric pressure and "gravitational pendulum assist" to connect cities on the East Coast of the US. They calculated that the run between Philadelphia and New York City would average 174 meters per second, that is 626 km/h (388 mph). When those plans were abandoned as too expensive, Lockheed engineer L.K. Edwards founded Tube Transit, Inc. to develop technology based on "gravity-vacuum transportation". In 1967 he proposed a Bay Area Gravity-Vacuum Transit for California that would run alongside the then-under construction BART system. It was never built.

21st Century

Research into trains running in partially-evacuated tubes is continuing.

Technical Characteristics

Modern systems (for smaller, i.e. "normal" tube diameters as used in the transport of small capsules) reach speeds of around 7.5 m (25 ft) per second, though some historical systems already achieved speeds of 10 m (33 ft) per second. Further, modern systems can also be computer-controlled, allowing, among other things, the tracking of any specific capsule. Varying air pressures also allow capsules to brake slowly, removing the jarring arrival that used to characterise earlier systems and make them unsuitable for fragile contents.

In Fiction

When pneumatic tubes first came into use in the 19th century, they symbolized technological progress and it was imagined that they would be common in the future. Jules Verne's *Paris in the Twentieth Century* (1863) includes suspended pneumatic tube trains that stretch across the oceans. Albert Robida's *The Twentieth Century* (1882) describes a 1950s Paris where tube trains have replaced railways, pneumatic mail is ubiquitous, and catering companies compete to deliver meals on tap to people's homes through pneumatic tubes. Edward Bellamy's *Looking Backward* (1888) envisions the world of 2000 as interlinked with tubes for delivering goods, while Michel Verne's *An Express of the Future* (1888) questions the sensibility of a transatlantic pneumatic subway. In Michel & Jules Verne's *The Day of an American Journalist in 2889* (1889) submarine tubes carry people faster than *aero-trains* and the *Society for Supplying Food to the Home* allows subscribers to receive meals pneumatically.

Later, because of their use by governments and large businesses, tubes began to symbolize bureaucracy. In George Orwell's *Nineteen Eighty-Four*, pneumatic tubes in the Ministry of Truth deliver newspapers to Winston's desk containing articles to be "rectified". Robert A. Heinlein's 1949 novella Gulf offered a more neutral view of their use in general postal delivery.)

In a sequence from the 1968 film *Baisers volés* (*Stolen Kisses*), Frannois Truffaut shows the fast transportation of a letter through the underground pneumatic tubes system in Paris. (This scene was later parodied in *The Simpsons* episode "Marge Gets a Job".)

In 1985, the movie *Brazil* also used tubes (as well as other anachronistic-seeming technologies) to evoke the stagnation of bureaucracy. At the start of each episode of the 1998 television series *Fantasy Island*, a darker version of the original, bookings for would-be visitors to the Island were sent to Mr. Roarke via a pneumatic tube from a dusty old travel agency.

The 1994 film version of *The Shadow* includes a sequence in which the camera follows a message capsule as it speeds through a pneumatic tube system. The implication is that the Shadow maintains a private network of tubes for the transportation of secret messages.

The failure of pneumatic tubes to live up to their potential as envisioned in previous centuries has placed them in the company of flying cars and dirigibles as ripe for ironic retro-futurism.

The 1960s cartoon series *The Jetsons* featured pneumatic tubes that people could step into and be sucked up and swiftly spat out at their destination. In the animated television series *Futurama*, set in the 31st century, large pneumatic tubes are used in cities for transporting people, whilst smaller ones are used to transport mail. The tubes in *Futurama* are also used to depict the endless confusion of bureaucracy: an immense network of pneumatic tubes connects all offices in New New York City to the "Central Bureaucracy", with all the capsules being deposited directly into a huge pile in the main filing room, with no sorting or organization.

In Ghostbusters II, the "river of slime" under New York city is found by the Ghostbusters boys to be flowing through an old pneumatic tube line-a reference to the Beach Pneumatic Transit tube.

In the American television show *Lost*, the *DHARMA* Initiative's Pearl research station has a pneumatic tube system. The character Locke put his drawing of the blast door map in the tube without a capsule. It was sucked up into the tube, indicating the system still functioned. The tube from the Pearl leads to a capsule dump.

In Kurt Vonnegut's *Slaughterhouse-Five* pneumatic tubes are used as a way to transport information from one place to the next when covering news articles.

In the popular video game *Bioshock*, pneumatic tubes are used to transport various items throughout the fictional city of Rapture.

Douglas Adams's 1998 computer game Starship Titanic features the "Succ-U-Bus" in almost every room-being a pneumatic pipe transport system which goes all around the ship; players must understand and use the Succ-U-Bus in order to progress and solve the puzzles.

Plastic Pressure Pipe Systems

Plastic pipework is used for the conveyance of drinking water, waste water, chemicals, heating and cooling fluids, foodstuffs, ultra-pure liquids, slurries, gases, compressed air and vacuum system applications.

Materials Used

- ABS (Acrylonitrile Butadiene Styrene)
- UPVC (Unplasticized Polyvinyl Chloride)
- CPVC (Post Chlorinated Polyvinyl Chloride)
- PP (Polypropylene)
- PE (Polyethylene) also known as LDPE, MDPE and HDPE (low, medium, and high density)
- PVDF (Polyvinylidene fluoride).

Material Characteristics

ABS (Acrylonitrile Butadiene Styrene)

ABS is used for the conveyance of potable water, slurries and chemicals.

UPVC (Unplasticized Polyvinyl Chloride) and CPVC (Post Chlorinated Polyvinyl Chloride)

UPVC has excellent chemical resistance across its operating temperature range, with a broad band of operating pressures. Due to its long-term strength characteristics, high stiffness and cost effectiveness, UPVC systems account for a large proportion of plastic piping installations. CPVC is resistant to many acids, bases, salts, paraffinic hydrocarbons, halogens and alcohols. It is not resistant to solvents, aromatics and some chlorinated hydrocarbons.

PP (Polypropylene)

Polypropylene is suitable for use with foodstuffs, potable and ultra pure waters, as well as within the pharmaceutical and chemical industries.

PE (Polyethylene)

Polyethylene has been successfully used for the safe conveyance of gases for many years.

PVDF (Polyvinylidene Fluoride)

PVDF has excellent chemical resistance which means that it is widely used in the chemical industry as a piping system for aggressive liquids.

Components of Plastic Pressure Pipe Systems

Pipes, fittings, valves, and accessories make up a plastic pressure pipe system. The range of pipe diameters for each pipe system does vary. However, the size ranges from 12 to 400 mm (0.472 to 15.748 in) and $^{3}D_{8}$ to 16 in (9.53 to 406.40 mm).

Pipes are extruded and are generally available in: 3 m (9.84 ft),4 m (13.12 ft), 5 m (16.40 ft), and 6 m (19.69 ft) straight lengths and 25 m (82.02 ft), 50 m (164.04 ft), 100 m (328.08 ft), and 200 m (656.17 ft) coils for LDPE and MDPE. Pipe fittings are moulded and come in many sizes: Tee 90° equal (straight and reducing), Tee 45°, Cross equal, Elbow 90° (straight and reducing), Elbow 45°, Short Radius Bend 90° Socket/coupler (straight and reducing), Union, End Caps, Reducing Bush, and Stub, Full face, and blanking flanges.

Valves are moulded and also come in many types: Ball Valve (also multiport), Butterfly valve, spring-, ball-, and swing-check non-return valves, Diaphragm valve, Knife Gate Valve, Globe Valve and Pressure Relief/Reduction Valves. Accessories are solvents, cleaners, glues, clips, backing rings, and gaskets.

Pig (Pipeline)

Pigging in the context of pipelines refers to the practice of using pipeline inspection gauges or 'pigs' to perform various maintenance operations on a pipeline. This is done without stopping the flow of the product in the pipeline.

Figure: *A cleaning pig for a 28-inch oil pipeline. The blue plastic disks seal against the inside of the pipe to propel the device and to remove loose sedimentation or scale buildup. The black rectangles at the top and the circular disks in the center are magnets to attract and remove any loose metal objects in the pipe.*

These operations include but are not limited to cleaning and inspecting of the pipeline. This is accomplished by inserting the pig into a 'pig launcher' (or 'launching station')-a funnel shaped Y section in the pipeline. The launcher/launching station is then closed and the pressure of the product in the pipeline is used to push it along down the pipe until it reaches the receiving trap-the 'pig catcher' (or receiving station).

If the pipeline contains butterfly valves, the pipeline cannot be pigged. Ball valves cause no problems because the inside diameter of the ball can be specified to be the same as that of the pipe (assuming they are full bore valves).

Figure: *A cleaning pig for a 6-inch oil pipeline. The wire brush encircles the shaft and scours the interior of the pipeline.*

Pigging has been used for many years to clean larger diameter pipelines in the oil industry. Today, however, the use of smaller diameter pigging systems is now increasing in many continuous and batch process plants as plant operators search for increased efficiencies and reduced costs.

Pigging can be used for almost any section of the transfer process between, for example, blending, storage or filling systems. Pigging systems are already installed in industries handling products as diverse as lubricating oils, paints, chemicals, toiletries, cosmetics and foodstuffs.

Pigs are used in lube oil or painting blending: they are used to clean the pipes to avoid cross-contamination, and to empty the pipes into the product tanks (or sometimes to send a component back to its tank). Usually pigging is done at the beginning and at the end of each batch, but sometimes it is done in the midst of a batch, e.g. when producing a premix that will be used as an intermediate component.

Pigs are also used in oil and gas pipelines: they are used to clean the pipes but also there are "smart pigs" used to measure things like pipe thickness and corrosion along the pipeline. They usually do not interrupt production, though some product can be lost when the pig is extracted. They can also be used to separate different products in a multiproduct pipeline.

Etymology

Pigs get their name from the squealing sound they make while travelling through a pipeline. (Disputed: 'PIG' is an acronym or backronym derived from the initial letters of the term 'Pipeline Inspection Gauge' or possibly 'Pipeline Inspection Gizmo' or 'Pipeline Internal Geometry' or 'Pipeline Inspection Gadget').

Pigging in Production Environments

Product and Time Saving

A major advantage of piggable systems is the potential resulting product savings. At the end of each product transfer, it is possible to clear out the entire line contents with the pig, either forwards towards the receipt point, or backwards to the source tank. There is no requirement for extensive line flushing.

Without the need for line flushing, pigging offers the additional advantage of a much more rapid and reliable product changeover. Product sampling at the receipt point becomes faster because the interface between products is very clear, and the old method of checking at intervals, until the product is on-specification, is considerably shortened.

Pigging can also be operated totally by a προγραμμεmable logic controller (PLC).

Environmental Issues

Pigging has a significant role to play in reducing the environmental impact of batch operations. Traditionally, the only way that an operator of a batch process could ensure a product was completely cleared from a line was to flush the line with a cleaning agent such as water or a solvent or even the next product. This cleaning agent then had to be subjected to effluent treatment or solvent recovery. If product was used to clear the line, the contaminated finished product was downgraded or dumped. In some cases, the finished product could contain polychlorinated biphenyl (PCB), which has been found to be carcinogenic. All of these problems can now be eliminated due to the very precise interface produced by modern pigging systems.

Safety Considerations

Pigging systems are designed so that the pig is loaded into the launcher, which is pressured up to launch the pig into the pipeline through a kicker line. In some cases, the pig is removed from the pipeline via the receiver at the end of each run. All systems must allow for the receipt of pigs at the launcher, as blockages in the pipeline may require the pigs to be pushed back to the launcher. Most of the time, systems are designed to pig the pipeline in either direction.

The pig is pushed either with an inert gas or a liquid; if pushed by gas, some systems can be adapted in the gas inlet in order to ensure pig's constant speed, whatever the pressure drop is. The pigs must be removed, as many pigs are rented, pigs wear and must be replaced, and cleaning pigs push contaminants from the pipeline such as wax, foreign objects, hydrates, etc., which must be removed from the pipeline. There are inherent risks in opening the barrel to atmosphere and care must be taken to ensure that the barrel is depressured prior to opening. If the barrel is not completely depressured, the pig can be ejected from the barrel and operators have been severely injured when standing in front of an open pig door. When the product is sour, the barrel should be evacuated to a flare system where the sour gas is burnt. Operators should be wearing a self-contained breathing apparatus when working on sour systems.

A few pigging systems utilize a "captive pig", and the pipeline is only opened up very occasionally to check the condition of the pig. At all other times, the pig is shuttled up and down the pipeline at the end of each transfer, and the pipeline itself is never opened up during process operation. These systems are not common.

Intelligent Pigging

Modern intelligent pigs are highly sophisticated instruments that vary in technology and complexity by the intended use and by manufacturer. An intelligent pig, or smart pig, includes electronics and sensors that collects various forms of data during the trip through the pipeline.

The electronics are sealed to prevent leakage of the pipeline product into the electronics since products can range from highly basic to highly acidic and can be of extremely high temperature. Many pigs use specific materials according to the product in the pipeline. Power for the electronics is provided by onboard batteries which are also sealed. Data recording may be by various means ranging from analog tape, digital tape, or solid state memory in more modern digital units.

Figure: *Inserting a pig into a natural gas pipeline*

The technology used to accomplish the service varies by the service required and the design of the pig, each pigging service provider may have unique and proprietary technologies to accomplish the service. Surface pitting and corrosion, as well as cracks and weld defects in steel/ferrous pipelines are often detected using magnetic flux leakage (MFL) pigs. Other "smart" pigs use electromagnetic acoustic transducers to detect pipe defects. Caliper pigs can measure the "roundness" of the pipeline to determine areas of crushing or other deformations. Some smart pigs can combine technologies such as MFL and Caliper into a single tool. Recent trials of pigs using acoustic resonance technology have been reported. During the pigging run the pig is unable to directly communicate with the outside world due to the distance underground or underwater and/or materials that the pipe is made of. For example, steel pipelines effectively prevent any reliable radio communications outside the pipe. It is therefore necessary that the pig use internal means to record its own movement during the trip. This may be done by gyroscope-assisted tilt sensors, odometers and other technologies. The pig will record this positional data so that the distance it moves along with any bends can be interpreted later to determine the exact path taken.

Location verification is often accomplished by surface instruments that record the pig's passage by either audible or gravinometric (or other) means. The sensors will record when they detect passage of the pig; this is then compared to the internal record for verification or adjustment. The external sensors may have GPS capability to assist in their location or even to transmit the pig's passage, but the pig itself usually cannot use GPS as it requires being able to receive the satellite signals.

After the pigging run has been completed, the positional data is combined with the pipeline evaluation data (corrosion, cracks, etc.) to provide a location-specific defect map and characterization. In other words, the combined data will tell the operator the location and type and size of each pipe defect. This is used to judge the severity of the defect and help repair crews locate and repair the defect quickly without having to dig up excessive amounts of pipeline. By evaluating the rate of change of a particular defect over several years, proactive plans can be made to repair the pipeline before any leakage or environmental damage occurs.

Pipeline Inspection Gauge

A pipeline inspection gauge or "PIG" in the pipeline industry is a tool that is sent down a pipeline and propelled by the pressure of the product in the pipeline itself. There are four main uses for pigs:

1. physical separation between different liquids being transported in pipelines;
2. internal cleaning of pipelines;
3. inspection of the condition of pipeline walls (also known as an Inline Inspection (ILI) tool);
4. capturing and recording geometric information relating to pipelines (e.g. size, position).

The original pigs were made from straw wrapped in wire used for cleaning. They made a squealing noise while travelling through the pipe, sounding to some like a pig squealing. The term "pipeline inspection gauge" was later created as a backronym.

One kind of pig is a soft, bullet shaped polyurethane foam plug that is forced through pipelines to separate products to reduce mixing. There are several types of pigs for cleaning. Some have tungsten studs or abrasive wire mesh on the outside to cut rust, scale, or paraffin deposits off the inside of the pipe. Others are plain plastic covered polyurethane.

Inline inspection pigs use various methods for inspecting a pipeline. A sizing pig uses one (or more) notched round metal plates that are used as gauges. The notches allow different parts of the plate to bend when a bore restriction is encountered. More complex systems exist for inspecting various aspects of the pipeline. Intelligent pigs, also called smart pigs, are used to inspect the pipeline with sensors and record the data for later analysis. These pigs use technologies such as MFL and ultrasonics to inspect the pipeline. Intelligent pigs may also use calipers to measure the inside geometry of the pipeline.

In 1961, the first intelligent pig was run by Shell Development. It demonstrated that a self contained electronic instrument could traverse a pipe line while measuring and recording wall thickness. The instrument used electromagnetic fields to sense wall integrity. In 1964 Tuboscope ran the first commercial instrument. It used MFL technology to inspect the bottom portion of the pipeline. The system used a black box similar to those used on aircraft to record the information.

A pig has been used as a plot device in three James Bond films: *Diamonds Are Forever*, where Bond disabled a pig to escape from a pipeline, *The Living Daylights*, where a pig was modified to secretly transport a person through the Iron Curtain, and *The World Is Not Enough*, where a pig was used to move a nuclear weapon through a pipeline.

A pig was also used as a plot device in the *Tony Hillerman* book *The Sinister Pig* where an abandoned pipeline from Mexico to the United States was used with a pig to transport illegal drugs.

Capacitive sensor probes are used in the process of detecting defects in polyethylene pipe gas pipeline. These probes are attached to the pig in which the pig is sent through the polyethylene pipe that will detect any defects in the outside of the pipe wall. This is done by using a triple plate capacitive sensor in which the electrostatic waves are propagated outware through the pipe's wall. Any change in dielectric material will result in a change in capacitance. Testing was conducted by NETL DOE research lab at the Battelle West Jefferson's Pipeline Simulation Facility (PSF) near Columbus, Ohio.

Trans-Alaska Pipeline System

The Trans Alaska Pipeline System (TAPS), includes the Trans Alaska Pipeline, 11 pump stations, several hundred miles of feeder pipelines, and the Valdez Marine Terminal. TAPS is one of the world's largest pipeline systems. It is commonly called the Alaska Pipeline, Trans Alaska Pipeline, or Alyeska Pipeline, (or the Pipeline as referred to in Alaska), but those terms technically apply only to the 800 miles (1,287 km) of the pipeline with the diameter of 48 inches (122 cm) that conveys oil from Prudhoc Bay, to Valdez, Alaska. The crude oil pipeline is privately owned by the Alyeska Pipeline Service Company.

The pipeline was built between 1974 and 1977 after the 1973 oil crisis caused a sharp rise in oil prices in the United States. This rise made exploration of the Prudhoe Bay oil field economically feasible. Environmental, legal, and political debates followed the discovery of oil at Prudhoe Bay in 1968, and the pipeline was built only after the oil crisis provoked the passage of legislation designed to remove legal challenges to the project.

The task of building the pipeline had to address a wide range of difficulties, stemming mainly from the extreme cold and the difficult, isolated terrain. The construction of the pipeline was one of the first large-scale projects to deal with problems caused by permafrost, and special construction techniques had to be developed to cope with the frozen ground. The project attracted tens of thousands of workers to Alaska, causing a boomtown atmosphere in Valdez, Fairbanks, and Anchorage.

The first barrel of oil traveled through the pipeline in 1977, and full-scale production began by the end of the year. Several notable incidents of oil leakage have occurred since, including those caused by sabotage, maintenance failures, and gunshot holes. The most significant oil spill associated with the pipeline was caused by the Exxon Valdez, and did not directly involve the pipeline. As of 2010, the pipeline has shipped almost 16 billion barrels (2.5×10^9 m^3) of oil.

Construction

Although the legal right-of-way was cleared by January 1974, cold weather, the need to hire workers, and construction of the Dalton Highway meant work on the pipeline itself did not begin until March. Between 1974 and July 22, 1977, when the first barrel of oil reached

Valdez, tens of thousands of people worked on the pipeline. Thousands of workers came to Alaska, attracted by the prospect of high-paying jobs at a time when most of the rest of the United States was undergoing a recession.

Construction workers endured long hours, cold temperatures, and brutal conditions. Difficult terrain, particularly in Atigun Pass, Keystone Canyon, and near the Sagavanirktok River forced workers to come up with solutions for unforeseen problems. Faulty welds and accusations of poor quality control caused a Congressional investigation that ultimately revealed little. More than $8 billion was spent to build the 800 miles (1,300 km) of pipeline, the Valdez Marine Terminal, and 12 pump stations. The construction effort also had a human toll. Thirty-two Alyeska and contract employees died from causes directly related to construction. That figure does not include common carrier casualties.

Impact

The construction of the Trans Alaska Pipeline System and its completion in 1977 had an immense effect on Alaska, the United States, and the rest of the world. Its impact has included economic, physical, and social repercussions running the gamut from life in small towns to the global oil market.

Boomtowns

Construction of the pipeline caused a massive economic boom in towns up and down the pipeline route. Prior to construction, most residents in towns like Fairbanks—still recovering from the devastating 1967 Fairbanks Flood—strongly supported the pipeline. By 1976, after the town's residents had endured a spike in crime, overstressed public infrastructure, and an influx of people unfamiliar with Alaska customs, 56 percent said the pipeline had changed Fairbanks for the worse. As great as the boom in Fairbanks was, it was proportionally more so in Valdez, where the population jumped from 1,350 in 1974 to 6,512 by the summer of 1975 and 8,253 in 1976.

This increase in population caused many adverse effects. Home prices skyrocketed—a home that sold for $40,000 in 1974 was purchased for $80,000 in 1975. In Valdez, lots of land that sold for $400 in the late 1960s went for $4,000 in 1973, $8,000 in 1974, and $10,000 in 1975. Home and apartment rentals were correspondingly squeezed upward by the rising prices and the demand from pipeline workers. Two-room log cabins with no plumbing rented for $500 per month. One two-bedroom home in Fairbanks housed 45 pipeline workers who shared beds on a rotating schedule for $40 per week. In Valdez, an apartment that rented for $286 per month in December 1974 cost $520 per month in March 1975 and $1,600 per month—plus two mandatory roommates—in April 1975. Hotel rooms were sold out as far away as Glenallen, 115 miles (185 km) north of Valdez.

The skyrocketing prices were driven by the high salaries paid to pipeline workers, who were eager to spend their money. The high salaries caused a corresponding demand for higher wages among non-pipeline workers in Alaska. Non-pipeline businesses often could not keep up with the demand for higher wages, and job turnover was high. Yellow Cab in Fairbanks had a turnover rate of 800 percent; a nearby restaurant had a turnover rate of more than 1,000 percent. Many positions were filled by high school students promoted above their experience level. To meet the demand, a Fairbanks high school ran in two shifts: one in the morning and the other in the afternoon in order to teach students who also worked eight hours per day. More wages and more people meant higher demand for goods and services. Waiting in line became a fact of life in Fairbanks, and the Fairbanks McDonalds became No. 2 in the world for sales—behind only the recently opened Stockholm store. Alyeska and its contractors bought in bulk from local stores, causing shortages of everything from cars to tractor parts, water softener salt, batteries and ladders.

The large sums of money being made and spent caused an upsurge in crime and illicit activity in towns along the pipeline route. This was exacerbated by the fact that police officers and state troopers resigned in large groups to become pipeline security guards at wages far in excess of those available in public-sector jobs. Fairbanks' Second Avenue became a notorious hangout for prostitutes, and dozens of bars operated throughout town. In 1975, the Fairbanks Police Department estimated between 40 and 175 prostitutes were working in the city of 15,000 people.

Prostitutes brought pimps, who then engaged in turf fights. In 1976, police responded to a shootout between warring pimps who wielded automatic firearms. By and large, however, the biggest police issue was the number of drunken brawls and fighting. On the pipeline itself, thievery was a major problem. Poor accounting and record keeping allowed large numbers of tools and large amounts of equipment to be stolen. The *Los Angeles Times* reported in 1975 that as many as 200 of Alyeska's 1,200 yellow-painted trucks were missing from Alaska and "scattered from Miami to Mexico City". Alyeska denied the problem and said only 20–30 trucks were missing. The theft problem was typified by pipeliners' practice of mailing empty boxes to pipeline camps. The boxes then would be filled with items and shipped out. After Alyeska ruled that all packages had to be sealed in the presence of a security guard, the number of packages being sent from camps dropped by 75 percent.

Economy of Alaska

The wealth generated by Prudhoe Bay and the other fields on the North Slope since 1977 is worth more than all the fish ever caught, all the furs ever trapped, all the trees chopped down; throw in all the copper, whalebone, natural gas, tin, silver, platinum, and anything else ever extracted from Alaska too. The balance sheet of Alaskan history is simple: One Prudhoe Bay is worth more in real dollars than everything that has been dug out, cut down, caught or killed in Alaska since the beginning of time.

Alaska Historian Terrence Cole

Since the completion of the Trans Alaska Pipeline System in 1977, the government of the state of Alaska has been reliant on taxes paid by oil producers and shippers. Prior to 1976, Alaska's personal income tax rate was 14.5 percent—the highest in the United States. The gross state product was $8 billion, and Alaskans earned $5 billion in personal income. Thirty years after the pipeline began operating, the state had no personal income tax, the gross state product was $39 billion, and Alaskans earned $25 billion in personal income. Alaska moved from the most heavily taxed state to the most tax-free state.

The difference was the Trans Alaska Pipeline System and the taxes and revenue it brought to Alaska. Alyeska and the oil companies injected billions of dollars into the Alaska economy during the construction effort and the years afterward. In addition, the taxes paid by those companies altered the tax structure of the state. By 1982, five years after the pipeline started transporting oil, 86.5 percent of Alaska revenue came directly from the petroleum industry.

The series of taxes levied on oil production in Alaska has changed several times since 1977, but the overall form remains mostly the same. Alaska receives royalties from oil production on state land. The state also has a property tax on oil production structures and transportation (pipeline) property—the only state property tax in Alaska. There is a special corporate income tax on petroleum companies, and the state taxes the amount of petroleum produced. This production tax is levied on the cost of oil at Pump Station 1. To calculate this tax, the state takes the market value of the oil, subtracts transportation costs (tanker and pipeline tariffs), subtracts production costs, then multiplies the resulting amount per barrel of oil produced each month. The state then takes a percentage of the dollar figure produced.

Under the latest taxation system, introduced by former governor Sarah Palin in 2007 and passed by the Alaska Legislature that year, the maximum tax rate on profits is 50 percent. The rate fluctuates based on the cost of oil, with lower prices incurring lower tax rates. The state also claims 12.5 percent of all oil produced in the state. This "royalty oil" is not taxed but is sold back to the oil companies, generating additional revenue. At a local level, the pipeline owners pay property taxes on the portions of the pipeline and the pipeline facilities that lay within districts that impose a property tax. This property tax is based on the pipeline's value (as assessed by the state) and the local property tax rate. In the Fairbanks North Star Borough, for example, pipeline owners paid $9.2 million in property taxes—approximately 10 percent of all property taxes paid in the borough.

The enormous amount of public revenue created by the pipeline provoked debates about what to do with the windfall. The record $900 million created by the Prudhoe Bay oil lease sale took place at a time when the entire state budget was less than $118 million, yet the entire amount created by the sale was used up by 1975. Taxes on the pipeline and oil carried by it promised to bring even more money into state coffers.

To ensure that oil revenue wasn't spent as it came in, the Alaska Legislature and governor Jay Hammond proposed the creation of an Alaska Permanent Fund—a long-term savings account for the state. This measure required a constitutional amendment, which was duly passed in November 1976. The amendment requires at least 25 percent of mineral extraction revenue to be deposited in the Permanent Fund. On February 28, 1977, the first deposit—$734,000—was put into the Permanent Fund. That deposit and subsequent ones were invested entirely in bonds, but debates quickly arose about the style of investments and what they should be used for.

In 1980, the Alaska Legislature created the Alaska Permanent Fund Corporation to manage the investments of the Permanent Fund, and it passed the Permanent Fund Dividend προγραμμε, which provided for annual payments to Alaskans from the interest earned by the fund. After two years of legal arguments about who should be eligible for payments, the first checks were distributed to Alaskans. After peaking at more than $40 billion in 2007, the fund's value declined to approximately $26 billion as of summer 2009. In addition to the Permanent Fund, the state also maintains the Constitutional Budget Reserve, a separate savings account established in 1990 after a legal dispute over pipeline tariffs generated a one-time payment of more than $1.5 billion from the oil companies. The Constitutional Budget reserve is run similar to the Permanent Fund, but money from it can be withdrawn to pay for the state's annual budget, unlike the Permanent Fund.

Oil Prices

Although the Trans-Alaska Pipeline System began pumping oil in 1977, it did not have a major immediate impact on global oil prices. This is partly because it took several years to reach full production and partly because U.S. production outside Alaska declined until the mid-1980s. The Iranian Revolution and OPEC price increases triggered the 1979 energy crisis despite TAPS production increases. Oil prices remained high until the late 1980s, when a stable international situation, the removal of price controls, and the peak of production at Prudhoe Bay contributed to the 1980s oil glut. In 1988, TAPS was delivering 25 percent of all U.S. oil production. As North Slope oil production declined, so did TAPS' share of U.S. production. Today, TAPS provides less than 17 percent of U.S. oil production.

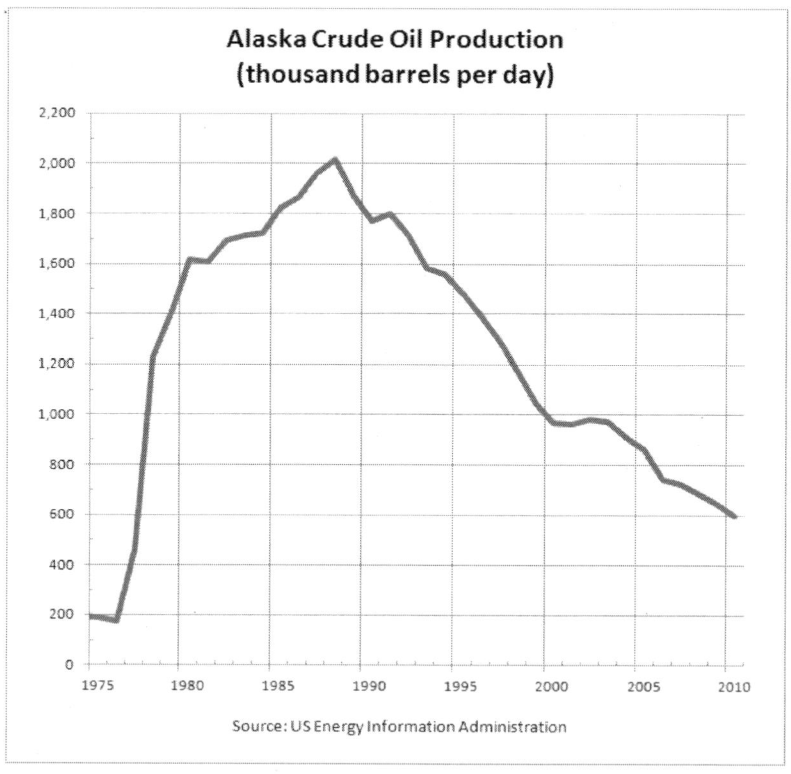

Figure: *Alaska oil production peaked in 1988.*

Social Impact

The pipeline attracts tens of thousands of visitors annually on pipeline tourism trips. Notable visitors have included Henry Kissinger, Jamie Farr, John Denver, President Gerald Ford, King Olav V of Norway, and Gladys Knight. Knight starred in one of two movies about the pipeline construction, *Pipe Dreams*. The other film was *Joyride*, and both were critically panned. Other films, such as *On Deadly Ground* and *30 Days of Night*, refer to the pipeline or use it as a plot device.

The pipeline has also inspired various forms of artwork. The most notable form of art unique to the pipeline are pipeline maps—portions of scrap pipe cut into the shape of Alaska with a piece of metal delineating the path of the pipeline through the map. Pipeline maps were frequently created by welders working on the pipeline, and the maps were frequently sold to tourists or given away as gifts. Other pipeline-inspired pieces of art include objects containing crude oil that has been transported through the pipeline.

Technical Details

Oil going into the Trans-Alaska Pipeline comes from one of several oil fields on Alaska's North Slope. The Prudhoe Bay Oil Field, the one most commonly associated with the pipeline, contributes oil, as do the Kuparuk, Alpine, Endicott, and Liberty oil fields, among others. Oil emerges from the ground at approximately 120 °F (49 °C) and cools to 111 °F (44 °C) by the time it reaches Pump Station 1 through feeder pipelines that stretch across the North Slope. North Slope crude oil has a specific gravity of 29.9 API at 60 °F (16 °C). In 2008, the pipeline carried approximately 700 thousand barrels per day (110,000 m^3/d), less than its theoretical maximum capacity of 2.14 million barrels per day (340,000 m^3/d) or its actual maximum of 2.03 million barrels per day (323,000 m^3/d) in 1988. From Pump Station 1 it takes an average of 11.9 days for oil to travel the entire length of the pipeline to Valdez, a speed of 3.7 miles per hour (6.0 km/h).

Pumping stations maintain the momentum of the oil as it goes through the pipeline. Pump Station 1 is the northernmost of 11 pump stations spread across the length of the pipeline. The original design called for 12 pump stations with 4 pumps each, but Pump Station 11 was never built. Nevertheless, the pump stations retained their intended naming system. Eight stations were operating at startup, and this number increased to 11 by 1980 as throughput rose. As of December 2006, only five stations were operating, with Pump Station 5 held in reserve. Pump Stations 2 and 7 have a capacity of moving

60,000 gallons/minute (227,125 l/min), while all other stations have a capacity of 20,000 gal/min (75,708 l/min). The pumps are natural-gas or liquid-fueled turbines.

Between the pipe stations and the end of the line is 800.3 miles (1,288.0 km) of 48-inch (1,200 mm) welded steel pipeline. The pipeline crosses 34 major streams or rivers and nearly 500 minor ones. Its highest point is at Atigun Pass, where the pipeline is 4,739 feet (1,444 m) above sea level. The maximum grade of the pipeline is 145%, at Thompson Pass in the Chugach Mountains. The pipeline was created in 40 and 60-foot (12.2 and 18.3-meter) sections. Forty-two thousand of these sections were welded together to make a double joint, which was laid in place on the line. Sixty-six thousand "field girth welds" were needed to join the double joints into a continuous pipeline. The pipe is of two different thicknesses: 466 miles (750 km) of it is 0.462 inches (1.17 cm) thick, while the remaining 334 miles (538 km) is 0.562 inches (1.43 cm) thick. More than 78,000 vertical support members hold up the aboveground sections of pipeline, and the pipeline contains 178 valves.

At the end of the pipeline is the Valdez Marine Terminal, which can store 9.18 million barrels (1,460,000 m^3) of oil. Eighteen storage tanks provide this capacity. They are 63.3 feet (19.3 m) tall and 250 feet (76 m) in diameter. They average 85% full at any given time—7.8 million barrels (1,240,000 m^3). Three power plants at the terminal generate 12.5 megawatts each. Four tanker berths are available for mooring ships in addition to two loading berths, where oil pumping takes place. More than 19,000 tankers have been filled by the marine terminal since 1977.

Maintenance

The pipeline is surveyed several times per day, mostly by air. Foot and road patrols also take place to check for problems such as leaks or pipe settling or shifting. The pipeline can be surveyed in just two hours, but most surveys take longer to ensure thoroughness. These external inspections are only part of standard maintenance, however. The majority of pipeline maintenance is done by pipeline pigs—mechanical devices sent through the pipeline to perform a variety of functions.

The most common pig is the scraper pig, which removes wax that precipitates out of the oil and collects on the walls of the pipeline. The colder the oil, the more wax buildup. This buildup can cause a variety of problems, so regular "piggings" are needed to keep the pipe clear. A second type of pig travels through the pipe and looks for corrosion.

Corrosion-detecting pigs use either magnetic or ultrasonic sensors. Magnetic sensors detect corrosion by analysing variations in the magnetic field of the pipeline's metal. Ultrasonic testing pigs detect corrosion by examining vibrations in the walls of the pipeline. Other types of pigs look for irregularities in the shape of the pipeline, such as if it is bending or buckling. "Smart" pigs, which contain a variety of sensors, can perform multiple tasks. Typically, these pigs are inserted at Prudhoe Bay and travel the length of the pipeline. In July 2009, a pig launcher was installed at Pump Station 8, near the midpoint of the pipeline. A third type of common maintenance is the installation and replacement of sacrificial anodes along the subterranean portions of pipeline. These anodes reduce the corrosion caused by electrochemical action that affect these interred sections of pipeline. Excavation and replacement of the anodes is required as they corrode.

Figure: *This scraper pig was retired from use in the pipeline and is on display.*

Incidents

The pipeline has at times been damaged due to sabotage, human error, maintenance failures, and natural disasters. By law, Alyeska is required to report significant oil spills to regulatory authorities. The Exxon Valdez oil spill is the best-known accident involving Alaska oil, but it did not involve the pipeline itself. Following the spill, Alyeska created a rapid response force that is paid for by the oil companies, including ExxonMobil, which was found liable for the spill.

The largest oil spill involving the main pipeline took place on February 15, 1978, when an unknown individual blew a 1-inch (2.54-centimeter) hole in it at Steele Creek, just east of Fairbanks.

Approximately 16,000 barrels (2,500 m^3) of oil leaked out of the hole before the pipeline was shut down. After more than 21 hours, it was restarted.

The steel pipe is resistant to gunshots and has resisted them on several occasions, but on October 4, 2001, a drunken gunman named Daniel Carson Lewis shot a hole into a weld near Livengood, causing the second-largest mainline oil spill in pipeline history. Approximately 6,144 barrels (976.8 m^3) leaked from the pipeline; 4,238 barrels (673.8 m^3) were recovered and reinjected into the pipeline. Nearly 2 acres (8,100 m^2) of tundra were soiled and were removed in the cleanup. The pipeline was repaired and was restarted more than 60 hours later. Lewis was found guilty in December 2002 of criminal mischief, assault, drunken driving, oil pollution, and misconduct. He was sentenced to 16 years in jail and ordered to repay the $17 million cleanup costs.

The pipeline was built to withstand earthquakes, forest fires, and other natural disasters. The 2002 Denali earthquake damaged some of the pipeline sliders designed to absorb similar quakes, and it caused the pipeline to shut down for more than 66 hours as a precaution. In 2004, wildfires overran portions of the pipeline, but it was not damaged and did not shut down.

In March 2006, corroded feeder pipelines on the North Slope gave way, spilling at least 6,310 barrels (1,003 m^3) of oil. In August 2006, during an inspection mandated by the United States Department of Transportation after the leak, severe corrosion was discovered. The transit pipelines were shut down for several days that month, and replacement of 16 miles (26 km) of transit pipeline began. The project was completed before Christmas Day 2008 at a cost of $500 million to BP.

In May 2010, as much as several thousands of barrels were spilled from a pump station near Fort Greely during a scheduled shutdown. A relief valve control circuit failed during a test of the fire control system, and oil poured into a tank and overflowed onto a secondary containment area.

A leak was discovered on Jan 8, 2011, in the basement of the booster pump at Pump Station 1. For more than 80 hours, pipeline flow was reduced to 5 percent of normal. An oil collection system was put in place, and full flow resumed until the pipeline was again shut down while a bypass was installed to avoid the leaking section.

Future of the Pipeline

By 2015, it is anticipated that daily oil throughput will approach 500,000 barrels per day (79,000 m^3/d), unless additional sources of oil are developed. As volumes decrease, Alaska will begin closing pump stations. The company intends to close all but four stations, because the lower throughput will require less pumping to maintain its momentum. While some reports supporting drilling in the ANWR coastal plain maintain that the pipeline may reach its minimum operating level of 200 thousand barrels per day (32,000 m^3/d) by 2020 the Trans-Alaska Pipeline System Renewal Environmental Impact Statement (TAPS Renewal EIS) estimated levels above this through at least 2032 due to ongoing exploration outside ANWR.

By law, Alaska is required to remove all trace of the pipeline after oil extraction is complete. No date has been set for this removal, but plans for it are being updated continuously.

Firestop

A firestop is a passive fire protection system of various components used to seal openings and joints in fire-resistance rated wall and/or floor assemblies, based on fire testing and certification listings.

Unprotected openings in fire separations void the fire-resistance ratings of the fire separations that contain them, allowing spread of fire past the limits of the fire safety plan of the entire building. Firestops are designed to restore the fire-resistance ratings of rated wall and/or floor assemblies by impeding the spread of fire through the opening by filling the openings with fire resistant materials.

Opening Types

- Electrical through-penetrations
- Mechanical through-penetrations
- Structural through-penetrations
- Unpenetrated openings (e.g. openings for future use)
- Re-entries of existing firestops
- Control or sway joints within fire-resistance rated wall or floor assemblies
- Junctions between fire-resistance rated wall or floor assemblies
- "Head-of-wall" (HOW) joints, where non-loadbearing wall assemblies meet floor assemblies,

Materials

Firestop components include Intumescents, Cementatious mortars, Silicone, Firestop pillows, Mineral fibres, and Rubber compounds.

Ratings

Firestops achieve a fire rating by combining certain materials in an arrangement specific to the item (e.g., pipe, cable) penetrating the fire rated wall or floor, and the construction arrangement of the fire rated wall or floor itself. The materials used in firestops do not have ratings on their own. For instance, a two-hour rated pipe penetration firestop may consist of a layer of caulking over packed rockwool. This overall arrangement provides the 2 hr. rating, not the caulking. However, both the individual firestop materials and the overall firestop assembly are 'listed' per listing and approval use and compliance.

Maintenance

Firestops should be maintained in accordance with the listing and approval use and compliance. Construction documentation sometimes includes an inventory of all firestops in a building, with drawings indicating location and the certification listings of each firestop. Using this documentation, the building owner can meet the requirements of the fire code related to fire barriers during the period of building occupation. Otherwise, improper repairs may result, which would violate the fire code, and could allow a fire to propagate between areas intended by code to be separated during a fire event.

Examples of Fire Barrier Penetrations Without Firestopping

Older buildings often have no firestops at all. In that case, a thorough inspection can identify all vertical and horizontal fire barriers and their fire ratings, and all breaches in these barriers, which can be sealed approved methods.

Firestops which are created by contractors or building maintenance personnel which are not listed are sometimes referred to as "deemed-to-comply", and are not credited with adequate fire resistance rating for building code compliance purposes. These are short term cost cutting measures at the expense of fire safety and code compliance. Common mistakes include citing a listing for products that may be for other uses. For instance, an insulation with an active listing for having a certain flamespread rating does not mean it is acceptable for firestopping purposes.

Re-entry

It is often necessary to the life of the building to install new electrical cables or mechanical systems piping through a hole in a fire rated barrier which has been properly fire stopped during initial construction, in which case the firestops are referred to as being "re-entered". In order to maintain the intent of the original fire protection plan of the building, firestops re-entries must be performed in compliance with the certification listing upon which the original installed configuration was based.

Work Sequencing with Spray Fireproofing

Spray fireproofing of structural steel is most efficiently accomplished before interior partitions are built, which can result in a conflict with fire stops in firewalls. Firestops must adhere to the bare, dry, unobstructed surfaces of the fire barrier which is penetrated (e.g., metal decking), or adjacent to an interface of two fire barriers at which fire stopping is required (e.g., beams). Spray fireproofing cannot be applied prior to fire stopping at these surfaces, as the fireproofing would obstruct adherence of the fire stop materials to the fire barrier surfaces. Spraying the upper room perimeter with fireproofing may also result in covering of wall/ceiling joints and through penetrations that require fire stopping. These joints are then not visually observed and not provided with proper fire stopping, violating the fire integrity of the passive fire barrier.

Tagging

Proper maintenance is enhanced by the installation of tags on each side of the firestop, containing the information necessary to reference to documents indicating the approved procedures for the original installation and re-entries. This requires knowledge of the exact certification listing that was used for each opening, be it a building joint or a penetrant through-penetration seal.

Testing and Certification

Acceptable certification listings include but are not limited to those available from;

- Underwriters Laboratories, or
- Underwriters Laboratories of Canada in North America, or
- Deutsches Intitut für Bautechnik in Germany, or
- Efectis in The Netherlands, France, Norway.
- FM Global provides testing and certification of firestops, and also certification of firetop contractors.

Regulations and Compliance

When the installed configuration is not in conformance with the appropriate certification listing, the fire-resistance rating may be less than expected. In those cases in which it is difficult to assess the impact, it often must be assumed to be zero, which means that the fire protection plan for the building is compromised. It is, therefore, necessary to be able to match each opening in fire-resistance rated wall or floor in a building with a certification listing. There are thousands of listings by various certification and testing laboratories. Both the Canadian and US Underwriters Laboratories each publish books containing just their own listings, including only those firestop manufacturers who have contracted with them for testing and certification.

Firestops should be routinely inspected and maintained to mitigate the effects of re-entries and time.

Trade Jurisdiction

In North American unionised construction sites, firestopping is often performed by members of the insulating trades. The insulators are the only building trade that includes firestop installation, theory and practical firestop training within its apprenticeship προγραμμε.

An agreement exists between the insulators and the electricians IBEW, which assigns firestop work from electricians to insulators, except that composite crews are required when working near live electrical conductors, whereby an electrician is required to observe and ensure the safety of the insulator.

Germany's GBA (Gütegemeinschaft Brandschutz im Ausbau) also offers a passive fire protection course, resulting in a certificate designation: "Brandschutzfachkraft" (~Passive Fire Protection Expert). In Europe as well as North America, all major firestop installers with nuclear installation experience are, by background, insulators first. The generic material types used and the skill sets needed between insulation and firestop installations are similar. Exceptions to the generic rule of thumb about firestopping being insulators' work, includes firestop devices that become an integral part of the plumbing system, which must be installed by plumbers during the forming of concrete.

Gasket

A gasket is a mechanical seal which fills the space between two or more mating surfaces, generally to prevent leakage from or into the joined objects while under compression. Gaskets allow *"less-than-perfect"* mating surfaces on machine parts where they can fill irregularities. Gaskets are commonly produced by cutting from sheet materials, such as gasket paper, rubber, silicone, metal, cork, felt, neoprene, nitrile rubber, fiberglass, Polytetrafluoroethylene (otherwise know as PTFE or Teflon) or a plastic polymer (such as polychlorotrifluoroethylene).

Gaskets for specific applications, such as high pressure steam systems, may contain asbestos. However, due to health hazards associated with asbestos exposure, non-asbestos gasket materials are used when practical.

It is usually desirable that the gasket be made from a material that is to some degree yielding such that it is able to deform and tightly fills the space it is designed for, including any slight irregularities. A few gaskets require an application of sealant directly to the gasket surface to function properly. Some (piping) gaskets are made entirely of metal and rely on a seating surface to accomplish the seal; the metal's own spring characteristics are utilized (up to but not passing σ_y, the material's yield strength). This is typical of some "ring joints" (RJ) or some other metal gasket systems such as those made by Graylock (an Oceaneering International company). These joints are known as R-con and E-con compressive type joints.

Properties

One of the more desirable properties of an effective gasket in industrial applications for compressed fiber gasket material is the ability to withstand high compressive loads. Most industrial gasket applications involve bolts exerting compression well into the 14 MPa (2000 psi) range or higher. Generally speaking, there are several truisms that allow for best gasket performance. One of the more tried and tested is: "The more compressive load exerted on the gasket, the longer it will last". There are several ways to measure a gasket material's ability to withstand compressive loading. The "hot compression test" is probably the most accepted of these tests. Most manufacturers of gasket materials will provide or publish the results of these tests. lens gaskets for high pressure

Gasket Design

Gaskets come in many different designs based on industrial usage, budget, chemical contact and physical parameters:

Sheet Gaskets

The premise is simple in that a sheet of material has the gasket shape "punched out" of it. This leads to a very crude, fast and cheap gasket. In older situations the material was compressed asbestos, but now generally a fibrous material such as graphite is used. These gaskets can fill many chemical requirements based on the inertness of the material used and fit many budgetary restraints. Common practice prevents these gaskets from being used in many industrial processes based on temperature and pressure concerns.

Solid Material Gaskets

The idea behind solid material is to use metals which cannot be punched out of sheets but are still cheap to produce. These gaskets generally have a much higher level of quality control than sheet gaskets and generally can withstand much higher temperatures and pressures. The key downside is that a solid metal must be greatly compressed in order to become flush with the flange head and prevent leakage. The material choice is more difficult; because metals are primarily used, process contamination and oxidation are risks. An additional downside is that the metal used must be softer than the flange — in order to ensure that the flange does not warp and thereby prevent sealing with future gaskets. Even so, these gaskets have found a niche in industry.

Spiral-Wound Gaskets

Spiral-wound gaskets comprise a mix of metallic and filler material. Generally, the gasket has a metal (normally carbon rich or stainless steel) wound outwards in a circular spiral (other shapes are possible) with the filler material (generally a flexible graphite) wound in the same manner but starting from the opposing side. This results in alternating layers of filler and metal. The filler material in these gaskets acts as the sealing element, with the metal providing structural support.

These gaskets have proven to be reliable in most applications, and allow lower clamping forces than solid gaskets, albeit with a higher cost.

Constant Seating Stress Gaskets

The constant seating stress gasket consists of two components; a solid carrier ring of a suitable material, such as stainless steel, and two sealing elements of some compressible material installed within two opposing channels, one channel on either side of the carrier ring. The sealing elements are typically made from a material (expanded graphite, expanded polytetraflouroethylene (PTFE), vermiculite, etc.) suitable to the process fluid and application. Constant seating stress gaskets derive their name from the fact that the carrier ring profile takes flange rotation (deflection under bolt preload) into consideration. With all other conventional gaskets, as the flange fasteners are tightened, the flange deflects radially under load, resulting in the greatest gasket compression, and highest gasket stress, at the outer gasket edge.

Since the carrier ring used in constant seating stress gaskets take this deflection into account when creating the carrier ring for a given flange size, pressure class, and material, the carrier ring profile can be adjusted to enable the gasket seating stress to be radially uniform across the entire sealing area. Further, because the sealing elements are fully confined by the flange faces in opposing channels on the carrier ring, any in-service compressive forces acting on the gasket are transmitted through the carrier ring and avoid any further compression of the sealing elements, thus maintaining a 'constant' gasket seating stress while in-service. Thus, the gasket is immune to common gasket failure modes that include creep relaxation, high system vibration, or system thermal cycles. The fundamental concept underlying the improved sealability for constant seating stress gaskets are that (i) if the flange sealing surfaces are capable of attaining a seal, (ii) the sealing elements are compatible with the process fluid and application, and (iii) the sufficient gasket seating stress is achieved on installation necessary to affect a seal, then the possibility of the gasket leaking in-service is greatly reduced or eliminated altogether.

The Patent for the constant seating stress gasket is held by JJENCO, Inc., Charlotte, North Carolina, USA. Various constant seating stress gasket designs for different applications are sold by that company under the tradename 'PerfectSeal'.

Double-Jacketed Gaskets

Double-jacketed gaskets are another combination of filler material and metallic materials. In this application, a tube with ends that resemble a "C" is made of the metal with an additional piece made to fit inside of the "C" making the tube thickest at the meeting points. The filler is pumped between the shell and piece. When in use the compressed gasket has a larger amount of metal at the two tips where contact is made (due to the shell/piece interaction) and these two places bear the burden of sealing the process. Since all that is needed is a shell and piece, these gaskets can be made from almost any material that can be made into a sheet and a filler can then be inserted. This is an effective option for most applications.

Kammprofile Gaskets

Kammprofile gaskets are used in many older seals since they have both a flexible nature and reliable performance. Kammprofiles work by having a solid corrugated core with a flexible covering layer. This arrangement allows for very high compression and an extremely tight seal along the ridges of the gasket. Since generally the graphite will fail instead of the metal core, Kammprofile can be repaired during later inactivity. Kammprofile has a high capital cost for most applications but this is countered by long life and increased reliability.

Flange Gasket

A flange gasket is a type of gasket made to fit between two sections of pipe that are flared to provide higher surface area.

Flange gaskets come in a variety of sizes and are categorized by their inside diameter and their outside diameter.

There are many standards in gasket for flanges of pipes. The gaskets for flanges can be divided in major 4 different categories:

1. Sheet gaskets
2. Corrugated metal gaskets
3. Ring gaskets
4. spiral wound gaskets.

Sheet gaskets are simple, they are cut to size either with bolt holes or without holes for standard sizes with various thickness and material suitable to media and temperature pressure of pipeline.

Ring gaskets also known as RTJ. They are mostly used in offshore oil-and gas pipelines and are designed to work under extremely high pressure. They are solid rings of metal in different cross sections like oval, round, octagonal etc. Sometimes they come with hole in center for pressure equalization.

Spiral wound gaskets are also used in high pressure pipelines and are made with stainless steel outer and inner rings and a center filled with spirally wound stainless steel tape wound together with graphite and Teflon, formed in V shape. Internal pressure acts upon the faces of the V, forcing the gasket to seal against the flange faces.

Improvements

Many gaskets contain minor improvements to increase lifespan or acceptable operating conditions:

- A common improvement is an inner compression ring. A compression ring allows for higher flange compression while preventing gasket failure. The effects of a compression ring are minimal and generally are just used when the standard design experiences a high rate of failure.
- A common improvement is an outer guiding ring. A guiding ring allows for easier installation and serves as a minor compression inhibitor. In some alkylation uses these can be modified on Double Jacketed gaskets to show when the first seal has failed through an inner lining system coupled with alkylation paint.

Ozone Cracking

Cracks can be formed in many different elastomers by ozone attack, and the characteristic form of attack of vulnerable rubbers is known as ozone cracking. The problem was formerly very common, especially in tires, but is now rarely seen in those products owing to preventive measures.

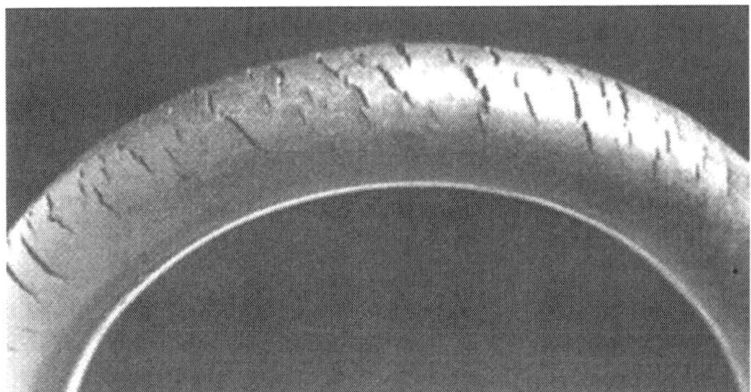

Figure: *Ozone cracking in natural rubber tubing*

However, it does occur in many other safety-critical items such as fuel lines and rubber seals, such as gaskets and O-rings, where ozone attack is considered unlikely. Only a trace amount of the gas is needed to initiate cracking, and so these items can also succumb to the problem.

Susceptible Elastomers

Tiny traces of ozone in the air will attack double bonds in rubber chains, with natural rubber, polybutadiene, styrene-butadiene rubber and nitrile rubber being most sensitive to degradation. Every repeat unit in the first three materials has a double bond, so every unit can be degraded by ozone. Nitrile rubber is a copolymer of butadiene and acrylonitrile units, but the proportion of acrylonitrile is usually lower than butadiene, so attack occurs. Butyl rubber is more resistant but still has a small number of double bonds in its chains, so attack is possible. Exposed surfaces are attacked first, the density of cracks varying with ozone gas concentration. The higher the concentration, the greater the number of cracks formed.

Ozone-resistant elastomers include EPDM, fluoroelastomers like Viton and polychloroprene rubbers like Neoprene. Attack is less likely because double bonds form a very small proportion of the chains, and with the latter, the chlorination reduces the electron density in the double bonds, therefore lowering their propensity to react with ozone. Silicone rubber, Hypalon and polyurethanes are also ozone-resistant.

Form of Cracking

Ozone cracks form in products under tension, but the critical strain is very small. The cracks are always oriented at right angles to the strain axis, so will form around the circumference in a rubber tube bent over. Such cracks are very dangerous when they occur in fuel pipes because the cracks will grow from the outside exposed surfaces into the bore of the pipe, so fuel leakage and fire may follow. Seals are also susceptible to attack, such as diaphragm seals in air lines. Such seals are often critical for the operation of pneumatic controls, and if a crack penetrates the seal, all functions of the system can be lost. Nitrile rubber seals are commonly used in pneumatic systems because of its oil resistance. However, if ozone gas is present, cracking will occur in the seals unless preventative measures are taken.

Figure: *Macrophotograph of ozone cracking in NBR (Nitrile Butadiene Rubber) diaphragm seal*

Ozone attack will occur at the most sensitive zones in a seal, especially sharp corners where the strain is greatest when the seal is flexing in use. The corners represent stress concentrations, so the tension is at a maximum when the diaphragm of the seal is bent under air pressure.

The seal shown at left failed from traces of ozone at circa 1 ppm, and once cracking had started, it continued as long as the gas was present. This particular failure led to loss of production on a semi-conductor fabrication line. The problem was solved by adding effective filters in the air line and by modifying the design to eliminate the very sharp corners. An ozone-resistant elastomer such as Viton was also considered as a replacement for the Nitrile rubber. The pictures were taken using ESEM for maximum resolution.

Ozonolysis

The reaction occurring between double bonds and ozone is known as ozonolysis when one molecule of the gas reacts with the double bond:

$$R_1R_2C{=}CR_3R_4 \xrightarrow{O_3} R_1R_2C{=}O + O{=}CR_3R_4$$

The immediate result is formation of an ozonide, which then decomposes rapidly so that the double bond is cleaved. This is the critical step in chain breakage when polymers are attacked. The strength of polymers depends on the chain molecular weight or degree of polymerization, the higher the chain length, the greater the mechanical strength (such as tensile strength). By cleaving the chain, the molecular weight drops rapidly and there comes a point when it has little strength whatsoever, and a crack forms. Further attack occurs in the freshly exposed crack surfaces and the crack grows steadily until it completes a circuit and the product separates or fails. In the case of a seal or a tube, failure occurs when the wall of the device is penetrated.

The carbonyl end groups which are formed are usually aldehydes or ketones, which can oxidise further to carboxylic acids. The net result is a high concentration of elemental oxygen on the crack surfaces, which can be detected using Energy-dispersive X-ray spectroscopy in the environmental SEM, or ESEM. The spectrum at left shows the high oxygen peak compared with a constant sulfur peak. The spectrum at right shows the unaffected elastomer surface spectrum, with a relatively low oxygen peak compared with the sulfur peak.

Prevention

The problem can be prevented by adding antiozonants to the rubber before vulcanization. Ozone cracks were commonly seen in automobile tire sidewalls, but are now seen rarely thanks to the use of these additives. A common and low cost antiozonant is a wax which bleeds to the surface and forms a protective layer, but other specialist chemicals are also widely used.

On the other hand, the problem does recur in unprotected products such as rubber tubing and seals, where ozone attack is thought to be impossible. Unfortunately, traces of ozone can turn up in the most unexpected situations. Using ozone-resistant rubbers is another way of inhibiting cracking.

For high value equipment where loss of function can cause serious problems, low cost seals may be replaced at frequent intervals so as to preclude failure. Ozone gas is produced during electric discharge by sparking or corona discharge for example. Static electricity can build up within machines like compressors with moving parts constructed from insulating materials. If those compressors feed pressurised air into a closed pneumatic system, then all seals in the system may be at risk from ozone cracking.

Ozone is also produced by the action of sunlight on volatile organic liquids or VOLs, such as gasoline vapour present in the air of towns and cities, in a problem known as photochemical smog. The ozone formed can drift many miles before it is destroyed by further reactions.

Polymer Degradation

Polymer degradation is a change in the properties—tensile strength, colour, shape, etc.—of a polymer or polymer-based product under the influence of one or more environmental factors such as heat, light or chemicals such as acids, alkalis and some salts. These changes are usually undesirable, such as cracking and chemical disintegration of products or, more rarely, desirable, as in biodegradation, or deliberately lowering the molecular weight of a polymer for recycling. The changes in properties are often termed "aging".

In a finished product such a change is to be prevented or delayed. Degradation can be useful for recycling/reusing the polymer waste to prevent or reduce environmental pollution. Degradation can also be induced deliberately to assist structure determination.

Polymeric molecules are very large (on the molecular scale), and their unique and useful properties are mainly a result of their size. Any loss in chain length lowers tensile strength and is a primary cause of premature cracking.

Commodity Oolymers

Today there are primarily seven commodity polymers in use: polyethylene, polypropylene, polyvinyl chloride, polyethylene terephthalate or PET, polystyrene, polycarbonate, and poly(methyl methacrylate) (Plexiglass). These make up nearly 98% of all polymers and plastics encountered in daily life. Each of these polymers has its own characteristic modes of degradation and resistances to heat, light and chemicals.

Polyethylene, polypropylene, and poly(methyl methacrylate) are sensitive to oxidation and UV radiation, while PVC may discolour at high temperatures due to loss of hydrogen chloride gas, and become very brittle. PET is sensitive to hydrolysis and attack by strong acids, while polycarbonate depolymerizes rapidly when exposed to strong alkalis. For example, polyethylene usually degrades by *random scission*-that is by a random breakage of the linkages (bonds) that hold the atoms of the polymer together. When this polymer is heated above 450 Celsius it becomes a complex mixture of molecules of various sizes that resemble gasoline. Other polymers-like polyalphamethylstyrene-undergo 'unspecific' chain scission with breakage occurring only at the ends; they literally unzip or depolymerize to become the constituent monomers.

Photoinduced Degradation

Most polymers can be degraded by photolysis to give lower molecular weight molecules. Electromagnetic waves with the energy of visible light or higher, such as ultraviolet light, X-rays and gamma rays are usually involved in such reactions.

Thermal Degradation

Chain-growth polymers like poly(methyl methacrylate) can be degraded by thermolysis at high temperatures to give monomers, oils, gases and water.

Solvolysis

Step-growth polymers like polyesters, polyamides and polycarbonates can be degraded by solvolysis and mainly hydrolysis to give lower molecular weight molecules. The hydrolysis takes place in the presence of water containing an acid or a base as catalyst. Polyamide is sensitive to degradation by acids and polyamide mouldings will crack when attacked by strong acids. For example, the fracture surface of a fuel connector showed the progressive growth of the crack from acid attack (Ch) to the final cusp (C) of polymer. The problem is known as stress corrosion cracking, and in this case was caused by hydrolysis of the polymer.

Oxidation

Polymers are susceptible to attack by atmospheric oxygen, especially at elevated temperatures encountered during processing to shape. Many process methods such as extrusion and injection moulding involve pumping molten polymer into tools, and the high temperatures needed for melting may result in oxidation unless precautions are taken. For example, a forearm crutch suddenly snapped and the user was severely injured in the resulting fall. The crutch had fractured across a polypropylene insert within the aluminium tube of the device, and infra-red spectroscopy of the material showed that it had oxidised, possible as a result of poor moulding.

Oxidation is usually relatively easy to detect owing to the strong absorption by the carbonyl group in the spectrum of polyolefins. Polypropylene has a relatively simple spectrum with few peaks at the carbonyl position (like polyethylene). Oxidation tends to start at tertiary carbon atoms because the free radicals formed here are more stable and longer lasting, making them more susceptible to attack by oxygen. The carbonyl group can be further oxidised to break the chain, this weakens the material by lowering its molecular weight, and cracks start to grow in the regions affected.

Galvanic Action

Polymer degradation by galvanic action was first described in the technical literature in 1990. This was the discovery that "plastics can corrode", i.e. polymer degradation may occur through galvanic action similar to that of metals under certain conditions. Normally, when two dissimilar metals such as copper (Cu) and iron (Fe) are put into contact and then immersed in salt water, the iron will undergo corrosion, or rust. This is called a galvanic circuit where the copper is the noble metal and the iron is the active metal, i.e., the copper is the cathode or positive (+) electrode and the iron is the anode, or negative (-) electrode. A battery is formed. It follows that plastics are made stronger by impregnating them with thin carbon fibers only a few micrometers in diameter known as carbon fiber reinforced polymers (CFRP). This is to produce materials that are high strength and resistant to high temperatures. The carbon fibers act as a noble metal similar to gold (Au) or platinum (Pt). When put into contact with a more active metal, for example with aluminium (Al) in salt water the aluminium corrodes. However in early 1990, it was reported that imide-linked resins in CFRP composites degrade when bare composite is coupled with an active metal in salt water environments. This is because corrosion not only occurs at the aluminium anode, but also at the carbon fiber cathode in the form of a very strong base with a pH of about 13. This strong base reacts with the polymer chain structure degrading the polymer. Polymers affected include bismaleimides (BMI), condensation polyimides, triazines, and blends thereof. Degradation occurs in the form of dissolved resin and loose fibers. The hydroxyl ions generated at the graphite cathode attack the O-C-N bond in the polyimide structure. Standard corrosion protection procedures were found to prevent polymer degradation under most conditions.

Chlorine-Induced Cracking

Another highly reactive gas is chlorine, which will attack susceptible polymers such as acetal resin and polybutylene pipework. There have been many examples of such pipes and acetal fittings failing in properties in the US as a result of chlorine-induced cracking. In essence, the gas attacks sensitive parts of the chain molecules (especially secondary, tertiary, or allylic carbon atoms), oxidizing the chains and ultimately causing chain cleavage. The root cause is traces of chlorine in the water supply, added for its anti-bacterial action, attack occurring even at parts per million traces of the dissolved gas. The chlorine attacks weak parts of a product, and in the case of an acetal resin junction in a water supply system, it is the thread roots that were attacked first, causing a brittle crack to grow. Discolouration on the fracture surface was caused by deposition of carbonates from the hard water supply, so the joint had been in a critical state for many months. The problems in the US also occurred to polybutylene pipework, and led to the material being removed from that market, although it is still used elsewhere in the world.

Biological Degradation

Biodegradable plastics can be biologically degraded by microorganisms to give lower molecular weight molecules. To degrade properly biodegradable polymers need to be treated like compost and not just left in a landfill site where degradation is very difficult due to the lack of oxygen and moisture.

Stabilisers

Hindered amine light stabilisers (HALS) stabilise against weathering by scavenging free radicals that are produced by photo-oxidation of the polymer matrix. UV-absorbers stabilises against weathering by absorbing ultraviolet light and converting it into heat. Antioxidants stabilize the polymer by terminating the chain reaction due to the absorption of UV light from sunlight. The chain reaction initiated by photo-oxidation leads to cessation of crosslinking of the polymers and degradation the property of polymers.

Vacuum Flange

A vacuum flange is a flange at the end of a tube used to connect vacuum chambers, tubing and vacuum pumps to each other.

Vacuum Flange Types

Several vacuum flange standards exist, and the same flange types are called by different names by different manufacturers and standards organizations.

KF/QF

The ISO standard quick release flange is known by the names Quick Flange (QF), Klein Flange (KF) or NW, sometimes also as DN.

The KF designation has been adopted by ISO, DIN, and Pneurop. KF flanges are made with a chamferred back surface that attached with a circular clamp and an elastomeric o-ring that is mounted in a metal centering ring. Standard sizes are indicated by the nominal inner diameter in millimetres for flanges 16 through 50 mm in diameter.

- DN16KF
- DN25KF
- DN40KF
- DN50KF.

ISO

The ISO large flange standard is known as LF, LFB, MF or sometimes just ISO flange. As in KF-flanges, the flanges are joined by a centering ring and an elastomeric o-ring. An extra spring-loaded circular clamp is often used around the large diameter o-rings to prevent them from rolling off from the centering ring during mounting.

The ISO large flanges come in two varieties. The ISO-K (or ISO LF) flanges are joined with double claw clamps which clamp to a circular groove on the tubing side of the flange. The ISO-F (or ISO LFB) flanges have holes for attaching the two flanges with bolts. Two tubes with ISO-K and ISO-F flanges can be joined together by clamping the ISO-K side with single claw clamps which are then bolted to the holes on the ISO-F side. ISO large flanges are available in sizes from 63 to 500 mm nominal tube diameter).

- DN63LF (63.5 mm)
- DN100LF (102 mm)
- DN160LF (160 mm)
- DN200LF (200 mm)
- DN250LF (254 mm)
- DN320LF (316 mm)
- DN400LF (400 mm)
- DN500LF (500 mm).

CF (Conflat)

CF (ConFlat) flanges use a copper gasket and knife-edge flange to achieve an ultrahigh vacuum seal. The term “ConFlat” is a registered trademark of Varian, Inc., so “CF” is commonly used by other flange manufacturers.

European, Asian size	*North American size [inches]*
DN10	1
DN16	1S‡ ("mini")
DN25	2[‡
DN40 (or: DN35)	2¾
DN50	3\‡
DN63	4½
DN75	4]‡
DN100	6
DN125	6¾
DN160 (or: DN150)	8
DN200	10
DN250	12
	13¼
	14
	16½

Each face of the two mating CF flanges has a knife edge which cuts into the softer metal gasket, providing an extremely leak-tight, metal-to-metal seal. Deformation of the metal gasket fills small defects in the flange, allowing Conflat flanges operate down to 10^{-13} torr (10^{-11} Pa) pressure. The gasket is partially recessed in a groove in each flange. The groove helps hold the gasket in place, which aligns the two flanges and also reduces gasket expansion during bake-out. For stainless steel conflat flanges baking temperatures of 450°C can be achieved; the temperature is limited by the choice of gasket material. CF flanges are sexless and interchangeable. North American flange sizes are given by flange outer diameter in inches while in Europe and Asia, sizes are given by tube inner diameter in millimetres:

ConFlat gaskets were originally invented by William Wheeler and other engineers at Varian in an attempt to build a flange that would not leak after baking.

Wheeler

A Wheeler flange is a large wire seal flange often used on large vacuum chambers.

ASA

ANSI has a flange standard called ASA. These flanges are elastomeric o-ring seal and can be used for both vacuum and pressure applications. Flange sizes are indicated by tube nominal inner diameter or by flange outer diameter (in inches): 1 (4.25 O.D.), 1.5 (5.00 O.D.), 2

(6.00 O.D.), 3 (7.50 O.D.), 4 (9.00 O.D.), 6 (11.00 O.D.), 8 (13.5 O.D.), 10 (16.00 O.D.).

Vacuum Gaskets

To achieve a vacuum seal, a gasket is required. An elastomeric o-ring gasket can be made of Buna rubber, viton fluoropolymer, silicon rubber or teflon. O-rings can be placed in a groove or may be used in combination with a centering ring or as a "captured" o-ring that is held in place by separate metal rings. Metal gaskets are used in ultra-high vacuum systems where the outgassing of the elastomer could be a significant gas load. A copper ring gasket is used with conflat flanges. Metal wire gaskets made of copper, gold or indium can be used.

Vacuum Feedthrough

A vacuum feedthrough is a flange that contains a vacuum-tight electrical, physical or mechanical connection to the vacuum chamber. An electrical feedthrough allows voltages to be applied to components under vacuum, for example a filament or heater. An example of a physical feedthrough is a vacuum tight connection for cooling water. A mechanical feedthrough is used for rotation and translation of components under vacuum. A wobble stick is a mechanical feedthrough device that can be used to pick up, move and otherwise manipulate objects in the vacuum chamber.

Washer (Mechanical)

A washer is a thin plate (typically disk-shaped) with a hole (typically in the middle) that is normally used to distribute the load of a threaded fastener, such as a screw or nut. Other uses are as a spacer, spring (belleville washer, wave washer), wear pad, preload indicating device, locking device, and to reduce vibration (rubber washer). Washers usually have an outer diameter (OD) about twice the length of their inner diameter (ID). Washers are usually metal or plastic. High quality bolted joints require hardened steel washers to prevent the loss of pre-load due to Brinelling after the torque is applied.

Rubber or fiber gaskets used in taps (or faucets, or valves) to stop the flow of water are sometimes referred to colloquially as *washers*; but, while they may look similar, washers and gaskets are usually designed for different functions and made differently. Washers are also important for preventing galvanic corrosion, particularly by insulating steel screws from aluminium surfaces. The origin of the word is unknown; the first recorded use of the word was in 1346, however the first time its definition was recorded was in 1611.

Types

Washers can be categorised into three types; *plain washers*, which spread a load, and prevent damage to the surface being fixed, or provide some sort of insulation such as electrical; *spring washers*, which have axial flexibility and are used to prevent fastening loosening due to vibrations; and *locking washers* which prevent fastening loosening by preventing unscrewing rotation of the fastening device; locking washers are usually also spring washers. The term washer is also often used for disc shaped devices used as grommets.

Plain Washers

A plain washer (or 'flat washer') is a flat annulus or ring, often of metal, used to spread the load of a screwed fastening. Additionally a plain washer may be used when the hole is a larger diameter than the fixing nut. A spherical washer is part of a self-aligning nut; it is a washer with one radiused surface, which is designed to be used in conjunction with a mating nut in order to correct for up to several degrees of misalignment between parts.

An anchor plate or wall washer is a large plate or washer connected to a tie rod or bolt. Anchor plates are used on exterior walls of masonry buildings, for structural reinforcement. Being visible, many anchor plates are made in a style that is decorative. A flange nut is a nut with an integral fixed washer. A torque washer is used in woodworking in combination with a carriage bolt; it has a square hole in the center where the carriage bolt square fits into. Teeth or prongs on the washer bite into the wood preventing the bolt from spinning freely when a nut is being tightened.

Spring and Locking Washers

Belleville washers, also known as a cupped spring washer or conical washer, has a slight conical shape, which provides an axial force when deformed. Curved disc spring is similar to a Belleville except the washer is curved in only one direction, therefore there are only four points of contact. Unlike Belleville washers, they only exert light pressures.

Wave washers have a "wave" in the axial direction, which provides spring pressure when compressed. Wave washers, of comparable size, do not produce as much force as Belleville washers. In Germany, they are used as a lock washer, however they are ineffective.

A split washer is a ring split at one point and bent into a helical shape. This causes the washer to exert a spring force between the fastener's head and the substrate. While this is supposed to act as a locking device, it is ineffective when it is tightened to become completely flat. Applicable standards are ASME B18.21.1, DIN 127B, and United

States Military Standard NASM 35338 (formerly MS 35338 and AN-935). A toothed lock washer, also known as a serrated washer or star washer, has serrations that extend radially inward and/or outward to bite into the bearing surface. This type of washer is effective as a lock washer when used with a soft substrate, such as aluminium or plastic. There are four types: internal, external, combination, and countersunk. The internal style have the serrations along the inner diameter of the washer, which makes them more aesthetically pleasing. The external style have the serrations around the outer diameter, which provides better holding power, because of the greater surface area. The combination style has serrations about both diameters, for maximum holding power. The countersunk style is designed to be used with flat-head screws.

Gaskets

The term washer is often applied to various gasket types such as those used to seal the control valve in taps. Crush washers are made of a soft metal such as aluminium or copper and are used to seal fluid or gas connections such as those found in an internal combustion engine. A shoulder washer is a plain washer type with integral cylindrical sleeve; they are used to keep separate different metal types, and a seals. This term is also used for electrically insulating grommets.

Specialised Types

A Keps nut or *K-lock nut* is a nut with an integral free spinning washer, assembly is easier because the washer is captive. A *top hat washer* is a shoulder washer type used in plumbing for tap fitting.

Materials

Common materials include steel, stainless steel, and plastic. *Hardened washers* are steel washers that have been heat treated. Other materials include aluminium, aramids, bimetals, bronze, brass, ceramics, copper, felt, fibers, hastelloy, iron, leather, mica, Inconel, Monel, rubber, silicon bronze, zinc, and titanium.

Standard Metric Flat Washers Sizes

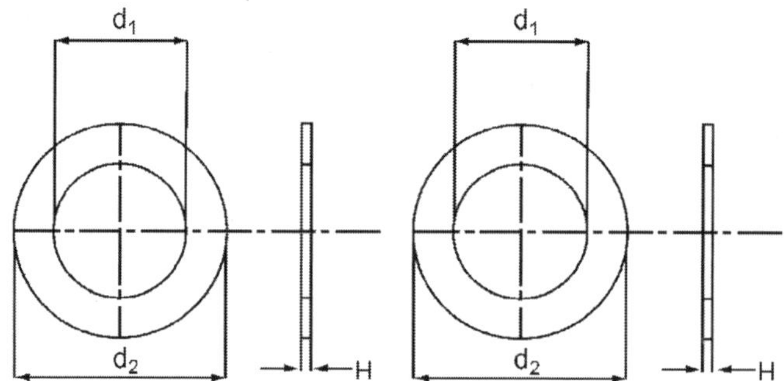

Nominal scre wsizes M (mm)		*Screw pitch thread P (mm)*		*Internal (Inside) diameter hole d1 (mm)*	*External (Outside) diameter d2 (mm)*	*thickness H (mm)*
1st choice	*2nd choice*	*coarse*	*fine*			
1		0.25		1.1	3	0.3
1.2		0.25		1.3	3.5	0.3
	1.4	0.3		1.5	4	0.3
1.6		0.35		1.7	4	0.3
	1.7			1.8	4.5	0.3
	1.8	0.35				
2		0.4		2.2	5	0.3
2.5		0.45		2.7	6	0.5
	2.6			2.8	7	0.5
3		0.5		3.2	7	0.5
	3.5	0.6		3.7	8	0.5
4		0.7		4.3	9	0.8
5		0.8		5.3	10	1
6		1	0.75	6.4	12	1.6
	7	1		7.4	14	1.6
8		1.25	1	8.4	16	1.6
10		1.5	1.25 or 1	10.5	20	2
12		1.75	1.5 or 1.25	13	24	2.5
	14	2	1.5	15	28	2.5
16		2	1.5	17	30	3
	18	2.5	2 or 1.5	19	34	3
20		2.5	2 or 1.5	21	37	3
	22	2.5	2 or 1.5	23	39	3
24		3	2	25	44	4
	27	3	2	28	50	4
30		3.5	2	31	56	4
	33	3.5	2	34	60	5
36		4	3	37	66	5
	39	4	3	40	72	6
42		4.5	3	43	78	7
	45	4.5	3	46	85	7
48		5	3	50	92	8
	52	5	4	54	98	8
56		5.5	4	58	105	9
	60	5.5	4	62	110	9
64		6	4	66	115	9
	68			70	120	10
72				74	125	10
	76			78	135	10
80				82	140	12
	85			87	145	12
90				93	160	12
	100			104	175	14

Hydrostatic Test

A hydrostatic test is a way in which leaks can be found in pressure vessels such as pipelines and plumbing. The test involves placing water, which is often dyed for visibility, in the pipe or vessel at the required pressure to ensure that it will not leak or be damaged. It is the most common method employed for testing pipes and vessels. Using this test helps maintain safety standards and durability of a vessel over time. Newly manufactured pieces are initially qualified using the hydrostatic test. They are then continually re-qualified at regular intervals using the *proof pressure test* which is also called the *modified hydrostatic test*. Hydrostatic testing is also a way in which a gas pressure vessel, such as a gas cylinder or a boiler, is checked for leaks or flaws. Testing is very important because such containers can explode if they fail when containing compressed gas.

Testing Procedures

Hydrostatic tests are conducted under the constraints of either the industry's or the customer's specifications. The vessel is filled with a nearly incompressible liquid-usually water or oil-and examined for leaks or permanent changes in shape. Red or fluorescent dyes are usually added to the water to make leaks easier to see. The test pressure is always considerably higher than the operating pressure to give a margin for safety. This margin of safety is typically 166.66% or 150% (in the case of many exemptions) of the designed pressure, depending on the regulations that apply. For example, if a cylinder was rated to DOT-2015 PSI (approximately 139 bar), it would be tested at around 3360 PSI (approximately 232 bar). Water is commonly used because it is almost incompressible (compressible only by weight, not air pressure), so will only expand by a very small amount should the vessel split. If high pressure gas were used, then the gas would expand to pV=nRT its compressed volume in an explosion, with the attendant risk of damage or injury. This is the risk which the testing is intended to mitigate. In other words using water is safer and takes less energy to do. Just one drop of water in every 5 seconds can cause a pressure change even up to 25 psi, while it would take a huge amount of gas to produce the same pressure change. Water is used mainly because the density of water is very close to 1.001 so using scale 1 g of water in the bowl would be 1 ml of cylinder expansion.

Small pressure vessels are normally tested using a water jacket test. The vessel is visually examined for defects and then placed in a container filled with water, and in which the change in volume of the vessel can be measured by monitoring the water level. For best accuracy, a digital scale is used to measure the smallest amounts of change. The vessel is then pressurized for a specified period, usually 30 or more seconds, and then depressurized again. The water level in the jacket is then examined. The level will be greater if the vessel being tested has been distorted by the pressure change and did not return to its original volume, or some of the pressurized water inside has leaked out. In both cases, this will normally signify that the vessel has failed the test. If the percentage of permanent expansion is more than 10%, or not up to DOT or customer standards, the cylinder fails, and then goes through a condemning process marking the cylinder as unsafe. This measures the overall leakage of a system instead of locating the leaks and additives can be added to the water to reduce resistivity and increase the sensitivity of the test. The hydrostatic test fluid can also clog small holes ($1x10^{-6}$ std cm^3/s or smaller) as a result of the increase in pressure. This is another reason why water is commonly used.

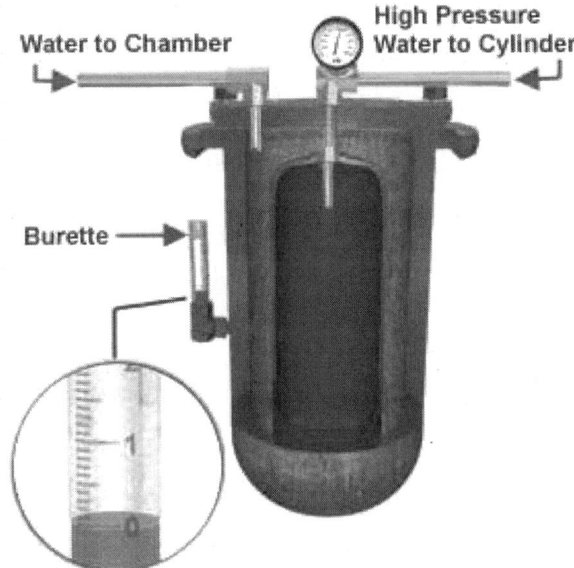

Figure: *Water Jacket Test*

All the information the tester needs is stamped onto the cylinder. This includes the DOT information, serial number, manufacturer, and manufacture date. Other information is stamped as needed such as the REE or how much the manufacturer specifies the cylinder should expand before it is considered unsafe. All this information is usually taken down and stored on a computer prior to the testing process. All this information is necessary for keeping track of when the cylinder has been or needs to be hydrotested. A simpler test, that is still considered a hydrostatic test but can be performed by anyone who has a garden hose, is to pressurize the vessel by filling it with water and to physically examine the outside for leaks. The pressure level achieved in this sort of test does not come close to the pressure level that would be used in a professional testing facility.

Examples

Portable fire extinguishers are safety tools that are required to be on hand in almost every public building. Fire extinguishers are also highly recommended in every home. Over time the conditions in which they are housed, and the manner in which they are handled have an impact on the structural integrity of the extinguisher. A structurally weakened fire extinguisher can malfunction or even burst when it is needed the most. To maintain the quality and safety of this product, hydrostatic testing must be utilized. All critical components of the fire extinguisher must be tested to ensure proper function. The cylinder would be tested by using the water jacket test.

As previously mentioned, the water pressure inside the tank will usually be 150% of the normal operating pressure. The change in volume of the cylinder is calculated by measuring the change in the water levels outside the cylinder. This can be done with a digital scale as well to detect the slightest changes, almost always in grams. The cylinder can also be visually checked for leaks or the pressure drop method can be utilized to measure the overall efficiency of the cylinder.

Pipeline Testing

Hydrotesting of pipe, pipelines and vessels is performed to expose defective materials that have missed prior detection, ensure that any remaining defects are insignificant enough to allow operation at design pressures, expose possible leaks and serve as a final validation of the integrity of the constructed system.

Buried high pressure oil and gas pipelines are tested for strength by pressurizing them to at least 125% of their maximum operating pressure (MAOP) at any point along their length. Since many long distance transmission pipelines are designed to have a steel hoop stress of 80% of specified minimum yield (SMYS) at MAOP, this means that the steel is stressed to SMYS and above during the testing, and test sections must be selected to ensure that excessive plastic deformation does not occur. Test pressures need not exceed a value that would produce a stress higher than yield stress at test temperature. Other codes require a more onerous approach. BS PD 8010-2 requires testing to 150% of the *design* pressure-which should not be less than the MAOP plus surge and other incidental effects that will occur during normal operation. Leak testing is performed by balancing changes in the measured pressure in the test section against the theoretical pressure changes calculated from changes in the measured temperature of the test section.

Testing Frequency

Most countries have legislation or building code that requires pressure vessels to be regularly tested, for example every two years (with a visual inspection annually) for high pressure gas cylinders and every five or ten years for lower pressure ones such as used in fire extinguishers. Gas cylinders which fail are normally destroyed as part of the testing protocol to avoid the dangers inherent in them being subsequently used. These common gas cylinders have the following requirements:

- DOT-3AL gas cylinders must be tested every 5 years and have an unlimited life.
- DOT-3HT gas cylinders must be tested every 3 years and have a 24 year life.
- SP-Cylinders Special Permit # http://hazmat.dot.gov/sp_app/special_permits/spec_perm_index.htm
- DOT-3AA gas cylinders must be tested every 5 years and have an unlimited life. (Unless stamped with a star (*) in which case the cylinder meets certain specifications and can have a 10 year hydrostatic test life)

In the U S and Canada, organizations such as ASTM, and ASME specify the guidelines for the different types of pressure vessels.

Pipe Network Aalysis

In fluid dynamics, pipe network analysis is the analysis of the fluid flow through a hydraulics network, containing several or many interconnected branches. The aim is to determine the flow rates and pressure drops in the individual sections of the network. This is a common problem in hydraulic design.

Description

In order to direct water to many individuals in a municipal water supply, many times the water is routed through a water supply network. A major part of this network may consist of interconnected pipes. This network creates a special class of problems in hydraulic design typically referred to as pipe network analysis. The modern solution for this is to use specialized software in order to automatically solve the problems. However, the problems can also be addressed with simpler methods like a spreadsheet equipped with a solver, or a modern graphing calculator.

Network Analysis

Once the friction factors are solved for, then we can start considering the network problem. We can solve the network by satisfying two conditions.

1. At any junction, the flow into a junction equals the flow out of the junction.
2. Between any two junctions, the head loss is independent of the path taken.

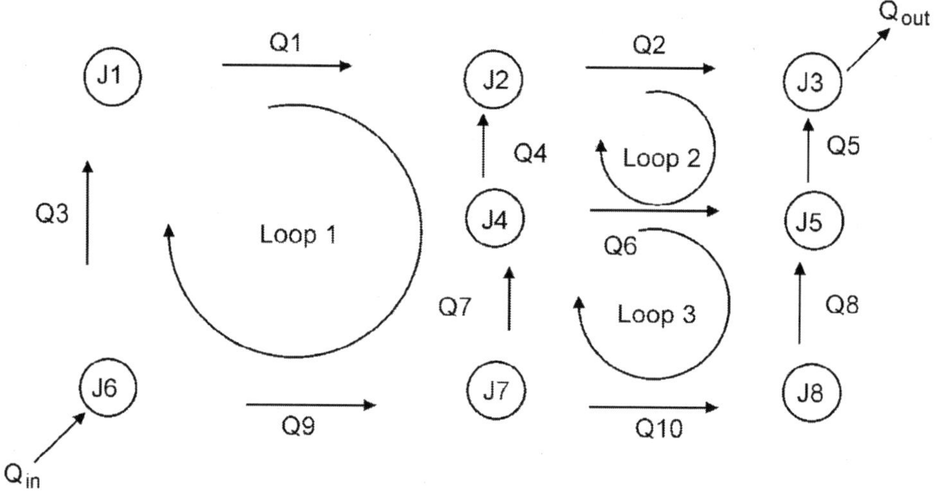

Figure 4

The classical approach for solving these networks is to use the Hardy Cross method. In this formulation, first you go through and create guess values for the flows in the network. That is, if Q7 enters a junction and Q6 and Q4 leave the same junction, then the initial guess must satisfy Q7 = Q6 + Q4.

After the initial guess is made, then, a loop is considered so that we can evaluate our second condition. Given a starting node, we work our way around the loop in a clockwise fashion, as illustrated by Loop 1. We add up the head losses according to the Darcy–Weisbach equation for each pipe if Q is in the same direction as our loop like Q1, and subtract the head loss if the flow is in the reverse direction, like Q4. In order to satisfy the second condition, we should end up with 0 about the loop if the network is completely solved. If the actual sum of our head loss is not equal to 0, then we will adjust all the flows in the loop by an amount given by the following formula, where a positive adjustment is in the clockwise direction.

$$\Delta Q = \frac{\sum \text{head loss}_c - \sum \text{head loss}_{cc}}{n \cdot (\sum \frac{\text{head loss}_c}{Q_c} + \sum \frac{\text{head loss}_{cc}}{Q_{cc}})},$$

where;

- n is 1.85 for Hazen-Williams and
- n is 2 for Darcy–Weisbach.

The clockwise specifier (c) means only the flows that are moving clockwise in our loop, while the counter-clockwise specifier (cc) is only the flows that are moving counter-clockwise. This adjustment won't solve the problem, since with most networks we will have several loops. It is ok to do this adjustment, however, because our flow changes won't alter condition 1, and therefore, our other loops will still satisfy condition 1. However, we should use the results from the first loop if we progress to any other loops. The more modern method is simply to create a set of conditions from your junctions and head-loss criteria. Then, use a Root-finding algorithm to find Q values that satisfy all the equations. The literal friction loss equations will use a term called Q^2, but we want to preserve any changes in direction. Create a separate equation for each loop where the head losses are added up, but instead of squaring Q, use $|Q| \cdot Q$ instead (with $|Q|$ the absolute value of Q) for the formulation so that any sign changes will reflect appropriately in the resulting head-loss calculation.

Chapter 5

Pipeline Protection

Cathodic protection (CP) is a technique used to control the corrosion of a metal surface by making it the cathode of an electrochemical cell. The simplest method to apply CP is by connecting the metal to be protected with another more easily corroded "sacrificial metal" to act as the anode of the electrochemical cell. Cathodic protection systems are used to protect a wide range of metallic structures in various environments. Common applications are; steel water or fuel pipelines and storage tanks; steel pier piles; ships and boats; offshore oil platforms and onshore oil well casings and metal reinforcement bars in concrete buildings and structures.

Cathodic protection can, in some cases, prevent stress corrosion cracking.

History

Cathodic protection was first described by Sir Humphry Davy in a series of papers presented to the Royal Society in London in 1824. After a series of tests, the first application was to the HMS Samarang in 1824. Sacrificial anodes made from iron were attached to the copper sheath of the hull below the waterline and dramatically reduced the corrosion rate of the copper. However, a side effect of the CP was to increase marine growth. Copper, when corroding, releases copper ions which have an anti-fouling effect. Since excess marine growth affected the performance of the ship, the Royal Navy decided that it was better to allow the copper to corrode and have the benefit of reduced marine growth, so CP was not used further.

Davy was assisted in his experiments by his pupil Michael Faraday, who continued his research after Davy's death. In 1834, he discovered the quantitative connection between corrosion weight loss and electric current and thus laid the foundation for the future application of cathodic protection.

Thomas Edison experimented with impressed current cathodic protection on ships in 1890, but was unsuccessful due to the lack of a suitable current source and anode materials. It would be 100 years after Davy's experiment before cathodic protection was used widely on oil pipelines in the United States.

Types

Galvanic CP

Galvanic anodes are designed and selected to have a more "active" voltage (more negative electrochemical potential) than the metal of the structure (typically steel). For effective CP, the potential of the steel surface is polarized (pushed) more negative until the surface has a uniform potential. At that stage, the driving force for the corrosion reaction is removed. The galvanic anode continues to corrode, consuming the anode material until eventually it must be replaced. The polarization is caused by the electron flow from the anode to the cathode. The driving force for the CP current is the difference in electrochemical potential between the anode and the cathode.

Metal	*Potential with respect to a Cu:CuSO$_4$ reference electrode in neutral pH environment (volts)*
Carbon, Graphite, Coke	+0.3
Platinum	0 to-0.1
Mill scale on Steel	-0.2
High Silicon Cast Iron	-0.2
Copper, brass, bronze	-0.2
Mild steel in concrete	-0.2
Lead	-0.5
Cast iron (not graphitized)	-0.5
Mild steel (rusted)	-0.2 to-0.5
Mild steel (clean)	-0.5 to-0.8
Commercially pure aluminium	-0.8
Aluminium alloy (5% zinc)	-1.05
Zinc	-1.1
Magnesium Alloy (6% Al, 3% Zn, 0.15% Mn)	-1.6
Commercially Pure Magnesium	-1.75

Galvanic or sacrificial anodes are made in various shapes and sizes using alloys of zinc, magnesium and aluminium. ASTM International publishes standards on the composition and manufacturing of galvanic anodes. In order for galvanic cathodic protection to work, the anode must possess a lower (that is, more negative) potential than that of the cathode (the structure to be protected). The table below shows a simplified galvanic series to show which metals can thus be combined.

Impressed Current CP

For larger structures, galvanic anodes cannot deliver economically enough current to provide complete protection. Impressed current cathodic protection (ICCP) systems use anodes connected to a DC power source. Usually this will be a cathodic protection rectifier, which converts an AC power supply to a DC output. In the absence of an AC supply, alternative power sources may be used, such as solar panels, wind power or gas powered thermoelectric generators

Anodes for ICCP systems are available in a variety of shapes and sizes. Common anodes are tubular and solid rod shapes or continuous ribbons of various materials. These include high silicon cast iron, graphite, mixed metal oxide, platinum and niobium coated wire and others.

Applications

Pipelines

Pipelines are routinely protected by a coating supplemented with cathodic protection. An ICCP system for a pipeline would consist of a DC power source, which is often an AC powered rectifier and an anode, or array of anodes buried in the ground (the anode groundbed).

The DC power source would typically have a DC output of between 10 and 50 amperes and 50 volts, but this depends on several factors, such as the size of the pipeline. The positive DC output terminal would be connected via cables to the anode array, while another cable would connect the negative terminal of the rectifier to the pipeline, preferably through junction boxes to allow measurements to be taken.

Anodes can be installed in a vertical hole and backfilled with conductive coke (a material that improves the performance and life of the anodes) or laid in a prepared trench, surrounded by conductive coke and backfilled. The choice of grounded type and size depends on the application, location and soil resistivity. The output of the DC source would then be adjusted to the optimum level after conducting various tests including measurements of electrochemical potential.

It is sometimes more economically viable to protect a pipeline using galvanic anodes. This is often the case on smaller diameter pipelines of limited length.

Ships

Cathodic protection on ships is often implemented by galvanic anodes attached to the hull, rather than using ICCP. Since ships are regularly removed from the water for inspections and maintenance, it is a simple task to replace the galvanic anodes. Galvanic anodes are generally shaped to reduced drag in the water and fitted flush to the hull to also try to minimize drag. Smaller vessels, with non-metallic hulls, such as yachts, will also use galvanic anodes to protect areas such as the rudder, but depend on an electrical connection between the anode and the item to be protected.

For ICCP on ships, a DC power supply is provided within the ship and the anodes mounted on the outside of the hull. The anode cables are introduced into the ship via a compression seal fitting and routed to the DC power source. The negative cable from the power supply is simply attached to the hull to complete the circuit. Ship ICCP anodes are flush-mounted, minimizing the effects of drag on the ship, and located a minimum 5 ft below the light load line in an area to avoid mechanical damage. The current density required for protection is a function of velocity and considered when selecting the current capacity and location of anode placement on the hull.

Some ships may require specialist treatment, for example aluminium hulls with steel fixtures will create an electrochemical cell where the aluminium hull can act as a galvanic anode and corrosion is enhanced. In cases like this, aluminium or zinc galvanic anodes can be used to offset the potential difference between the aluminium hull and the steel fixture. If the steel fixtures are large, several galvanic anodes may be required, or even a small ICCP system.

Marine

Marine CP covers many areas, jetties, harbours, offshore structures. The variety of different types of structure leads to a variety of systems to provide protection. Typically, galvanic anodes are favoured, but ICCP can also often be used.

Steel in Concrete

The application to concrete reinforcement is slightly different in that the anodes and reference electrodes are usually embedded in the concrete at the time of construction when the concrete is being poured.

The usual technique for concrete buildings, bridges and similar structures is to use ICCP, but there are systems available that use the principle of galvanic CP as well, although in the UK at least, the use of galvanic anodes for atmospherically exposed reinforced concrete structures is considered experimental.

For ICCP, the principle is the same as any other ICCP system. However, in a typical atmospherically exposed concrete structure such as a bridge, there will be many more anodes distributed through the structure as opposed to an array of anodes as used on a pipeline. This makes for a more complicated system and usually an automatically controlled DC power source is used, possibly with an option for remote monitoring and operation. For buried or submerged structures, the treatment is similar to that of any other buried or submerged structure.

Galvanic systems offer the advantage of being easier to fit retrospectively, since the anodes are fitted on the concrete surface and do not need any control systems as ICCP does.

For pipelines constructed from pre-stressed concrete cylinder pipe (PCCP), the techniques used for CP are generally as for steel pipelines except that there is a need to take steps to avoid an excessive level of potential that can produce cause possible damage to the prestressing wire

> *The steel wire in a PCCP pipeline is stressed to the point that any corrosion of the wire can result in failure. An additional problem is that any excessive hydrogen ions as a result of an excessively negative potential can cause hydrogen embrittlement of the wire, also resulting in failure. The failure of too many wires will result in catastrophic failure of the PCCP. To implement ICCP therefore requires very careful control to ensure satisfactory protection. A simpler option is to use galvanic anodes, which are self limiting and need no control.*

Internal CP

Vessels, pipelines and tanks which are used to store or transport liquids can also be protected from corrosion on their internal surfaces by the use of cathodic protection. ICCP and galvanic systems can be used.

Galvanized Steel

Galvanizing generally refers to hot-dip galvanizing which is a way of coating steel with a layer of metallic zinc. Galvanized coatings are quite durable in most environments because they combine the barrier properties of a coating with some of the benefits of cathodic protection. If the zinc coating is scratched or otherwise locally damaged and steel is exposed, the surrounding areas of zinc coating form a galvanic cell with the exposed steel and protect it from corrosion. This is a form of localized cathodic protection-the zinc acts as a sacrificial anode.

It should be noted that galvanizing, while using the principle of cathodic protection, is not actually cathodic protection. CP requires the anode to be separate from the metal surface to be protected, with an ionic connection through the electrolyte and an electron connection through a connecting cable, bolt or similar. This means that any area of the protected structure within the electrolyte can be protected, whereas in the case of galvanizing, only areas very close to the zinc are protected. Hence, a larger area of bare steel would only be protected around the edges.

Testing

Electrochemical corrosion potential is measured with reference electrodes. Copper-copper(II) sulfate electrodes are used for structures in contact with soil or fresh water. Silver chloride electrodes or Saturated calomel electrodes (SCE) are used for seawater applications. The methods are described in along with the sources of error in the voltage that appears on the display of the meter. Interpretation of electrochemical corrosion potential measurements to determine the potential at the interface between the anode of the corrosion cell and the electrolyte requires training and cannot be expected to match the accuracy of measurements done in laboratory work. It is widely recognized that errors can be introduced in all measurements where there is a lack of training.

Problems

Production of Hydrogen Ions: A side effect of improperly applied cathodic protection is the production of hydrogen ions, leading to its absorption in the protected metal and subsequent hydrogen embrittlement of welds and materials with high hardness. Under normal conditions, the ionic hydrogen will combine at the metal surface to create hydrogen gas, which cannot penetrate the metal. Hydrogen ions, however, are small enough to pass through the crystalline steel structure, and lead in some cases to hydrogen embrittlement.

Cathodic Disbonding: This is a process of disbondment of protective coatings from the protected structure (cathode) due to the formation of hydrogen ions over the surface of the protected material (cathode). Disbonding can be exacerbated by an increase in alkali ions and an increase in cathodic polarization. The degree of disbonding is also reliant on the type of coating, with some coatings affected more than others. Cathodic protection systems should be operated so that the structure does not become excessively polarized, since this also promotes disbonding due to excessively negative potentials. Cathodic disbonding occurs rapidly in pipelines that contain hot fluids because the process is accelerated by heat flow.

Cathodic Shielding: Effectiveness of cathodic protection systems on steel pipelines can be impaired by the use of solid film backed dielectric coatings such as polyethylene tapes, shrinkable pipeline sleeves, and factory applied single or multiple solid film coatings. This phenomenon occurs because of the high electrical resistivity of these film backings. Protective electric current from the cathodic protection system is blocked or shielded from reaching the underlying metal by the highly resistive film backing. Cathodic shielding was first defined in the 1980s as being a problem, and technical papers on the subject have been regularly published since then. A 1999 report concerning a 20,600 bbl (3,280 m^3). spill from a Saskatchewan crude oil line contains an excellent definition of the cathodic shielding problem:

"The triple situation of disbondment of the (corrosion) coating, the dielectric nature of the coating and the unique electrochemical environment established under the exterior coating, which acts as a shield to the electrical CP current, is referred to as CP shielding. The combination of tenting and disbondment permits a corrosive environment around the outside of the pipe to enter into the void between the exterior coating and the pipe surface. With the development of this CP shielding phenomenon, impressed current from the CP system cannot access exposed metal under the exterior coating to protect the pipe surface from the consequences of an aggressive corrosive environment. The CP shielding phenomenon induces changes in the potential gradient of the CP system across the exterior coating, which are further pronounced in areas of insufficient or sub-standard CP current emanating from the pipeline's CP system. This produces an area on the pipeline of insufficient CP defense against metal loss aggravated by an exterior corrosive environment."

Cathodic shielding is referenced in a number of the standards listed below. Newly issued USDOT regulation Title 49 CFR 192.112, in the section for *Additional design requirements for steel pipe using alternative maximum allowable operating pressure* requires that "The pipe must be protected against external corrosion by a non-shielding coating". Also, the NACE SP0169:2007 standard defines shielding in section 2, cautions against the use of materials that create electrical shielding in section 4.2.3, cautions against use of external coatings that create electrical shielding in section 5.1.2.3, and instructs readers to take 'appropriate action' when the effects of electrical shielding of CP current are detected on an operating pipeline in section 10.9.

Standards

- 49 CFR 192.112-Requirements for Corrosion Control-Transportation of natural and other gas by pipeline: minimum federal safety standards
- ASME B31Q 0001-0191
- ASTM G 8, G 42-Evaluating Cathodic Disbondment resistance of coatings
- DNV-RP-B401-Cathodic Protection Design-Det Norske Veritas
- EN 12068:1999-Cathodic protection. External organic coatings for the corrosion protection of buried or immersed steel pipelines used in conjunction with cathodic protection. Tapes and shrinkable materials
- EN 12473:2000-General principles of cathodic protection in sea water
- EN 12474:2001-Cathodic protection for submarine pipelines
- EN 12495:2000-Cathodic protection for fixed steel offshore structures
- EN 12499:2003-Internal cathodic protection of metallic structures
- EN 12696:2000-Cathodic protection of steel in concrete
- EN 12954:2001-Cathodic protection of buried or immersed metallic structures. General principles and application for pipelines
- EN 13173:2001-Cathodic protection for steel offshore floating structures
- EN 13174:2001-Cathodic protection for harbor installations
- EN 13509:2003-Cathodic protection measurement techniques
- EN 13636:2004-Cathodic protection of buried metallic tanks and related piping
- EN 14505:2005-Cathodic protection of complex structures
- EN 15112:2006-External cathodic protection of well casing
- EN 50162:2004-Protection against corrosion by stray current

from direct current systems

- BS 7361-1:1991-Cathodic Protection
- NACE SP0169:2007-Control of External Corrosion on Underground or Submerged Metallic Piping Systems
- NACE TM 0497-Measurement Techniques Related to Criteria for Cathodic Protection on Underground or Submerged Metallic Piping Systems.

Anodic Protection

Anodic protection (AP) is a technique to control the corrosion of a metal surface by making it the anode of an electrochemical cell and controlling the electrode potential in a zone where the metal is passive. AP is used to protect metals that exhibit passivation in environments whereby the current density in the freely corroding state is significantly higher than the current density in the passive state over a wide range of potentials.

Anodic protection is used for carbon steel storage tanks containing extreme pH environments including concentrated sulfuric acid and 50 percent caustic soda where cathodic protection is not suitable due to very high current requirements.

An anodic protection system includes an external power supply connected to auxiliary cathodes and controlled by a feedback from reference electrode. Careful design and control is required when using anodic protection for several reasons, including excessive current when passivation is lost or unstable, leading to possible accelerated corrosion.

Sacrificial Metal

A sacrificial metal is a metal used as a sacrificial anode in cathodic protection that corrodes to prevent a primary metal from corrosion, galvanization or rusting.

Equation

When two metals touch each other and water is present, electrolysis occurs. One well known example is the reaction between zinc (Zn) and iron (Fe). Zn atoms ionize as it is more electropositive and is oxidized and corrodes.

Zn(s)’™Zn^{2+}(aq) +2e (oxidation).

Uses

Sacrificial metals are widely used to prevent other metals from rusting, for example food cans. Most of the food can is coated with a layer of metal that is more electropositive than the metal (mostly iron) inside the food can, preventing the iron contaminating the food with Fe^{2+} ions.

Valve

A valve is a device that regulates, directs or controls the flow of a fluid (gases, liquids, fluidized solids, or slurries) by opening, closing, or partially obstructing various passageways. Valves are technically pipe fittings, but are usually discussed as a separate category. In an open valve, fluid flows in a direction from higher pressure to lower pressure.

The simplest, and very ancient, valve is simply a freely hinged flap which drops to obstruct fluid (gas or liquid) flow in one direction, but is pushed open by flow in the opposite direction.

Figure: *These water valves are operated by handles.*

Valves are used in a variety of contexts, including industrial, military, commercial, residential, and transport. The industries in which the majority of valves are used are oil and gas, power generation, mining, water reticulation, sewerage and chemical manufacturing.

In daily life, most noticeable are plumbing valves, such as taps for tap water. Other familiar examples include gas control valves on cookers, small valves fitted to washing machines and dishwashers, safety devices fitted to hot water systems, and valves in car engines. In nature, veins acting as valves are controlling the blood circulation; heart valves control the flow of blood in the chambers of the heart and maintain the correct pumping action. Valves play a vital role in industrial applications ranging from transportation of drinking water to control of ignition in a rocket engine.

Valves may be operated manually, either by a handle, lever or pedal. Valves may also be automatic, driven by changes in pressure, temperature, or flow. These changes may act upon a diaphragm or a piston which in turn activates the valve, examples of this type of valve found commonly are safety valves fitted to hot water systems or boilers.

More complex control systems using valves requiring automatic control based on an external input (i.e., regulating flow through a pipe to a changing set point) require an actuator. An actuator will stroke the valve depending on its input and set-up, allowing the valve to be positioned accurately, and allowing control over a variety of requirements.

Applications

Valves vary widely in form and application. Sizes typically range from 0.1 mm to 60 cm. Special valves can have a diameter exceeding 5 meters.

Valve cost ranges from simple inexpensive disposable valves to specialized valves cost thousands of US dollars per inch of diameter.

Disposable valves may be found inside common household items including mini-pump dispensers and aerosol cans.

Types

Valves are quite diverse and may be classified into a number of basic types. Valves may also be classified by how they are actuated:

Hydraulic

Hydraulics is a topic in applied science and engineering dealing with the mechanical properties of liquids. Fluid mechanics provides the theoretical foundation for hydraulics, which focuses on the engineering uses of fluid properties. In fluid power, hydraulics is used for the generation, control, and transmission of power by the use of pressurized liquids. Hydraulic topics range through most science and engineering disciplines, and cover concepts such as pipe flow, dam design, fluidics and fluid control circuitry, pumps, turbines, hydropower, computational fluid dynamics, flow measurement, river channel behaviour and erosion.

Free surface hydraulics is the branch of hydraulics dealing with free surface flow, such as occurring in rivers, canals, lakes, estuaries and seas. Its sub-field open channel flow studies the flow in open channels.

Basic Types

Valves can be categorized into the following basic types:

- Ball valve, for on/off control without pressure drop, and ideal for quick shut-off since a 90° turn offers complete shut-off angle, compared to multiple turns required on most manual valves.

A ball valve is a valve with a spherical disc, the part of the valve which controls the flow through it. The sphere has a hole, or port, through the middle so that when the port is in line with both ends of the valve, flow will occur. When the valve is closed, the hole is perpendicular to the ends of the valve, and flow is blocked. The handle or lever will be inline with the port position letting you "see" the valve's position. The ball valve, along with the butterfly valve and plug valve, are part of the family of *quarter turn valves*.

Ball valves are durable and usually work to achieve perfect shutoff even after years of disuse. They are therefore an excellent choice for shutoff applications (and are often preferred to globe valves and gate valves for this purpose). They do not offer the fine control that may be necessary in throttling applications but are sometimes used for this purpose.

Ball valves are used extensively in industrial applications because they are very versatile, supporting pressures up to 1000 bars and temperatures up to 250°C. Sizes typically range from 0.5 cm to 30 cm. They are easy to repair and operate.

The body of ball valves may be made of metal, plastic or metal with a ceramic center. The ball is often chrome plated to make it more durable. A ball-check valve is a type of check valve with a ball without a hole for a disc, and is not covered in this chapter.

The genericized trademark *ball-o-fix* is occasionally used after the original UK market leader.

Types of Ball Valve

There are five general body styles of ball valves: *single body*, *three piece body*, *split body*, *top entry*, and *welded*. The difference is based on how the pieces of the valve—especially the casing that contains the ball itself—are manufactured and assembled. The valve operation is the same in each case.

In addition, there are different styles related to the bore of the ball mechanism itself:

- A *full port* or more commonly known *full bore* ball valve has an over-sized ball so that the hole in the ball is the same size as the pipeline resulting in lower friction loss. Flow is unrestricted but the valve is larger and more expensive so this is only used where free flow is required, for example in pipelines which require pigging.

- In *reduced port* or more commonly known *reduced bore* ball valves, flow through the valve is one pipe size smaller than the valve's pipe size resulting in flow area being smaller than pipe. As the flow discharge remains constant and is equal to area of flow (A) times velocity (V), $A_1V_1 = A_2V_2$ the velocity increases with reduced area of flow.
- A *V port* ball valve has either a 'v' shaped ball or a 'v' shaped seat. This allows the orifice to be opened and closed in a more controlled manner with a closer to linear flow characteristic. When the valve is in the closed position and opening is commenced the small end of the 'v' is opened first allowing stable flow control during this stage. This type of design requires a generally more robust construction due to higher velocities of the fluids, which might damage a standard valve.
- A trunnion ball valve has additional mechanical anchoring of the ball at the top and the bottom, suitable for larger and higher pressure valves (say, above 10 cm and 40 bars).
- Cavity filler Ball Valve. Many industries encounter problem with residues in the ball valve. Where the fluid is meant for human consumption, residues may also be health hazard, and when where the fluid changes from time to time contamination of one fluid with another may occur. Residues arise because in the half open position of the ball valve a gap is created between the ball bore and the body in which fluid can be trapped. To avoid the fluid getting into this cavity, the cavity has to be plugged, which can be done by extending the seats in such a manner that it is always in contact with the ball. This type of ball valve is known as Cavity Filler Ball Valve.

Manually operated ball valves can be closed quickly and thus there is a danger of water hammer. Some ball valves are equipped with an actuator that may be pneumatically or motor operated. These valves can be used either for on/off or flow control. A pneumatic flow control valve is also equipped with a positioner which transforms the control signal into actuator position and valve opening accordingly.

Three-Way and Four-Way Ball Valves

Three-way ball valves have an L-or T-shaped hole through the middle. It is easy to see that a T valve can connect any pair of ports, or all three, together, but the 45 degree position which might disconnect all three leaves no margin for error. The L valve can connect the center port to either side port, or disconnect all three, but it cannot connect the side ports together.

Multi-port ball valves with 4 ways, or more, are also commercially available, the inlet way often being orthogonal to the plane of the outlets. For special applications, such as driving air-powered motors from forward to reverse, the operation is performed by rotating a single lever four-way valve. The 4-way ball valve has two L-shaped ports in the ball that do not interconnect, sometimes referred to as an "×" port.

Ball valves in sizes up to 2 inch generally come in single piece, two or three piece designs. One piece ball valves are almost always reduced bore, are relatively inexpensive and generally are throw-away. Two piece ball valves are generally slightly reduced (or standard) bore, they can be either throw-away or repairable. The 3 piece design allows for the center part of the valve containing the ball, stem & seats to be easily removed from the pipeline. This facilitates efficient cleaning of deposited sediments, replacement of seats and gland packings, polishing out of small scratches on the ball, all this without removing the pipes from the valve body. The design concept of a three piece valve is for it to be repairable.

Butterfly Valve

A butterfly valve is a valve which can be used for isolating or regulating flow. The closing mechanism takes the form of a disk. Operation is similar to that of a ball valve, which allows for quick shut off. Butterfly valves are generally favoured because they are lower in cost to other valve designs as well as being lighter in weight, meaning less support is required. The disc is positioned in the center of the pipe, passing through the disc is a rod connected to an actuator on the outside of the valve. Rotating the actuator turns the disc either parallel or perpendicular to the flow. Unlike a ball valve, the disc is always present within the flow, therefore a pressure drop is always induced in the flow, regardless of valve position.

A butterfly valve is from a family of valves called quarter-turn valves. The "butterfly" is a metal disc mounted on a rod. When the valve is closed, the disc is turned so that it completely blocks off the passageway. When the valve is fully open, the disc is rotated a quarter turn so that it allows an almost unrestricted passage of the fluid. The valve may also be opened incrementally to throttle flow.

***Figure:** Large butterfly valve used on a hydroelectric power station water inlet pipe in Japan.*

There are different kinds of butterfly valves, each adapted for different pressures and different usage. The resilient butterfly valve, which uses the flexibility of rubber, has the lowest pressure rating. The high performance butterfly valve, used in slightly higher-pressure systems, features a slight offset in the way the disc is positioned, which increases the valve's sealing ability and decreases its tendency to wear. The valve best suited for high-pressure systems is the triple offset butterfly valve, which makes use of a metal seat, and is therefore able to withstand a greater amount of pressure.

Structure

Butterfly valves are valves with a circular body and a rotary motion disk closure member which is pivotally supported by its stem. A butterfly valve can appear in various styles, including eccentric and high-performance valves. These are normally a type of valve that uses a flat plate to control the flow of water. As well as this, butterfly valves are used on firefighting apparatus and typically are used on larger lines, such as front and rear suction ports and tank to pump lines. A butterfly valve is also a type of flow control device, used to make a fluid start or stop flowing through a section of pipe. The valve is similar in operation to a ball valve. Rotating the handle turns the plate either parallel or perpendicular to the flow of water, shutting off the flow. It is a very well known and well used design.

Types

1. 'Concentric Butterfly Valves'-This type of valves has a resilient rubber seat with a metal disc.
2. 'Double Eccentric Butterfly Valves'-This type of valves are also referred as 'High Performance Butterfly Valves' or 'Double Offset Butterfly Valves'. Different type of materials is used for seat and disc.
3. 'Triple Eccentric Butterfly Valves'-This type of valves are also commonly called 'Triple Offset Butterfly Valves'. The seats are either laminated or solid metal seat design.

Figure: *Duplex valve in wafer butterfly configuration.*

Triple Eccentric(Offset) Design

Triple eccentric(offset) Design prevents galling and scratches between the metal seat and the metal disc due to its unique design. The only time where the seal comes into contact with the seat is at the point of complete closure. Triple offset valves are generally used in applications which require bi-directional tight shut-off in Oil & Gas, LNG/NPG terminal and tanks, Chemical Factories, Shipbuilding. Widely use for dirty/heavy oil to prevent extrusion.

Wafer-Style Butterfly Valves

The wafer style butterfly valve is designed to maintain a seal against bi-directional pressure differential to prevent backflow in systems designed for unidirectional flow. It accomplishes this with a tightly fitting seal, *i.e.*, gasket, o-ring, precision machined, and a flat valve face on the upstream and downstream sides of the valve.

Lug-style butterfly valve

Lug-style valves have threaded inserts at both sides of the valve body. This allows them to be installed into a system using two sets of bolts and no nuts. The valve is installed between two flanges using a separate set of bolts for each flange. This setup permits either side of the piping system to be disconnected without disturbing the other side.

A lug-style butterfly valve used in dead end service generally has a reduced pressure rating. For example a lug-style butterfly valve mounted between two flanges has a 150 psi pressure rating. The same valve mounted with one flange, in dead end service, has a 75 psi rating.

Check Valve

A check valve, clack valve, non-return valve or one-way valve is a mechanical device, a valve, which normally allows fluid (liquid or gas) to flow through it in only one direction. Check valves are two-port valves, meaning they have two openings in the body, one for fluid to enter and the other for fluid to leave. There are various types of check valves used in a wide variety of applications. Check valves are often part of common household items. Although they are available in a wide range of sizes and costs, check valves generally are very small, simple, and/or inexpensive. Check valves work automatically and most are not controlled by a person or any external control; accordingly, most do not have any valve handle or stem. The bodies (external shells) of most check valves are made of plastic or metal.

Figure: *Tilting disc inconel check valve*

An important concept in check valves is the cracking pressure which is the minimum upstream pressure at which the valve will operate. Typically the check valve is designed for and can therefore be specified for a specific cracking pressure.

Heart valves are essentially inlet and outlet check valves for the heart ventricles, since the ventricles act as pumps.

Types of Check Valves

Backwater valve (for sanitary drainage system) protects lower located rooms against flooding caused by return flow of sewage waters. Such risk occurs most often in sanitary drainage systems connected to combined sewerage systems and in rainwater drainage systems. It may be caused by intense rainfall, thaw or flood. Backwater valve prevents rats and other rodents entering the sanitary and rainwater drainage systems and, consequently, the building interiors. It protects also against unpleasant smells in case of longer breaks in system use.

A ball check valve is a check valve in which the closing member, the movable part to block the flow, is a spherical ball. In some ball check valves, the ball is spring-loaded to help keep it shut. For those designs without a spring, reverse flow is required to move the ball toward the seat and create a seal. The interior surface of the main seats of ball check valves are more or less conically-tapered to guide the ball into the seat and form a positive seal when stopping reverse flow.

Ball check valves are often very small, simple, and cheap. They are commonly used in liquid or gel minipump dispenser spigots, spray devices, some rubber bulbs for pumping air, etc., manual air pumps and some other pumps, and refillable dispensing syringes. Although the balls are most often made of metal, they can be made of other materials, or in some specialized cases out of artificial ruby. High pressure HPLC pumps and similar applications commonly use small inlet and outlet ball check valves with both balls and seats made of artificial ruby, for both hardness and chemical resistance. After prolonged use, such check valves can eventually wear out or the seat can develop a crack, requiring replacement. Therefore, such valves are made to be replaceable, sometimes placed in a small plastic body tightly-fitted inside a metal fitting which can withstand high pressure and which is screwed into the pump head.

There are similar check valves where the disc is not a ball, but some other shape, such as a poppet energized by a spring. Ball check valves should not be confused with ball valves, which is a different type of valve in which a ball acts as a controllable rotor to stop or direct flow.

A diaphragm check valve uses a flexing rubber diaphragm positioned to create a normally-closed valve. Pressure on the upstream side must be greater than the pressure on the downstream side by a certain amount, known as the pressure differential, for the check valve to open allowing flow. Once positive pressure stops, the diaphragm automatically flexes back to its original closed position.

A swing check valve or tilting disc check valve is check valve in which the disc, the movable part to block the flow, swings on a hinge or trunnion, either onto the seat to block reverse flow or off the seat to allow forward flow. The seat opening cross-section may be perpendicular to the centerline between the two ports or at an angle. Although swing check valves can come in various sizes, large check valves are often swing check valves. The flapper valve in a flush-toilet mechanism is an example of this type of valve. Tank pressure holding it closed is overcome by manual lift of the flapper. It then remains open until the tank drains and the flapper falls due to gravity. Another variation of this mechanism is the clapper valve, used in applications such firefighting and fire life safety systems. A hinged gate only remains open in the inflowing direction. The clapper valve often also has a spring that keeps the gate shut when there is no forward pressure. A stop-check valve is a check valve with override control to stop flow regardless of flow direction or pressure. In addition to closing in response to backflow or insufficient forward pressure (normal check-valve behaviour), it can also be deliberately shut by an external mechanism, thereby preventing any flow regardless of forward pressure.

A lift-check valve is a check valve in which the disc, sometimes called a *lift*, can be lifted up off its seat by higher pressure of inlet or upstream fluid to allow flow to the outlet or downstream side. A guide keeps motion of the disc on a vertical line, so the valve can later reseat properly. When the pressure is no longer higher, gravity or higher downstream pressure will cause the disc to lower onto its seat, shutting the valve to stop reverse flow.

A duckbill valve is a check valve in which flow proceeds through a soft tube that protrudes into the downstream side. Back-pressure collapses this tube, cutting off flow.

Multiple check valves can be connected in series. For example, a double check valve is often used as a backflow prevention device to keep potentially contaminated water from siphoning back into municipal water supply lines. There are also *double ball check valves* in which there are two ball/seat combinations sequentially in the same body to ensure positive leak-tight shutoff when blocking reverse flow;

and piston check valves, wafer check valves, and ball-and-cone check valves.

Applications

Pumps: Check valves are often used with some types of pumps. Piston-driven and diaphragm pumps such as metering pumps and pumps for chromatography commonly use inlet and outlet ball check valves. These valves often look like small cylinders attached to the pump head on the inlet and outlet lines. Many similar pump-like mechanisms for moving volumes of fluids around use check valves such as ball check valves. The feed pumps or injectors which supply water to steam boilers are fitted with check valves to prevent back-flow.

Figure: *The check valves on this steam locomotive are located under the small cover between the chimney and the main dome*

Industrial Processes

Check valves are used in many fluid systems such as those in chemical and power plants, and in many other industrial processes.

Check valves are also often used when multiple gases are mixed into one gas stream. A check valve is installed on each of the individual gas streams to prevent mixing of the gases in the original source. For example, if a fuel and an oxidizer are to be mixed, then check valves will normally be used on both the fuel and oxidizer sources to ensure that the original gas cylinders remain pure and therefore nonflammable.

Domestic Use

Some types of irrigation sprinklers and drip irrigation emitters have small check valves built into them to keep the lines from draining when the system is shut off. Also used with most home made snowmakers.

Check valves used in domestic heating systems to prevent vertical convection, especially in combination with solar thermal installations, also are called gravity brake. Rainwater harvesting systems that are plumbed into the main water supply of a utility provider may be required to have one or more check values fitted to prevent contamination of the primary supply by rainwater.

Choke Valve

A Choke valve is a type of valve designed to create choked flow in a fluid. Over a wide range of valve settings the flow through the valve can be understood by ignoring the viscosity of the fluid passing through the valve; the rate of flow is determined only by the ambient pressure on the upstream side of the valve.

In automotive contexts, a choke valve modifies the air pressure in the intake manifold of an internal combustion engine, thereby altering the ratio of fuel and air quantity entering the engine. Choke valves are generally used in naturally aspirated engines with carburetors, to supply a richer fuel mixture when starting the engine. Most choke valves in engines are butterfly valves mounted in the manifold above the carburetor jet, to produce a higher partial vacuum and thereby draw more fuel into the intake stream.

In heavy industrial or fluid engineering contexts, a choke valve is a particular design of valve that raises and lowers a solid cylinder (called a "plug" or "stem") which is placed around or inside another cylinder that has holes or slots. The design of a choke valve means fluids flowing through the cage are coming from all sides and that the streams of flow (through the holes or slots) collide with each other at the center of the cage cylinder, thereby dissipating the energy of the fluid through "flow impingement". The main advantage of choke valves is that they can be designed to be totally linear in their flow rate.

Automotive

A choke valve is sometimes installed in the carburetor of internal combustion engines. Its purpose is to restrict the flow of air, thereby enriching the fuel-air mixture while starting the engine. Depending on engine design and application, the valve can be activated manually by the operator of the engine (via a lever or pull handle) or automatically by a temperature-sensitive mechanism called an autochoke. Choke valves are important for normally-aspirated gasoline engines because small droplets of gasoline do not evaporate well within a cold engine. By restricting the flow of air into the throat of the carburetor, the choke valve reduces the pressure inside the throat, which causes a proportionally greater amount of fuel to be pushed from the main jet into the combustion chamber during cold-running operation. Once the engine is warm (from combustion), opening the choke valve restores the carburetor to normal operation, supplying fuel and air in the correct stoichiometric ratio for clean, efficient combustion.

Note that the term "choke" is applied to the carburetor's enrichment device even when it works by a totally different method. Commonly SU carburetors have "chokes" that work by lowering the fuel jet to a narrower part of the needle. Some others work by introducing an additional fuel route to the constant depression chamber.

Chokes were nearly universal in automobiles until fuel injection began to supplant carburetors. Choke valves are still common in other internal-combustion applications, including most small portable engines, motorcycles, small propeller-driven airplanes, riding lawn mowers, and normally-aspirated marine engines.

Industrial

Heavy duty industrial choke valves control the flow to a certain Flow Coefficient (Cv) determined by how far the valve is opened. They are regularly used in the oil industry and for highly erosive and corrosive purposes, they are often made of tungsten carbide or inconel.

Diaphragm Valve

Diaphragm valves (or membrane valves) consists of a valve body with two or more ports, a diaphragm, and a "saddle" or seat upon which the diaphragm closes the valve. The valve is constructed from either plastic or steel.

Originally, the diaphragm valve was developed for use in non-hygienic applications. Later on the design was adapted for use in the bio-pharmaceutical industry by using compliant materials that can withstand sanitizing and sterilizing methods.

There are two main categories of diaphragm valves: one type seals over a "weir" (saddle) and the other (sometimes called a "straight-way" valve) seals over a seat. The main difference is that a saddle-type valve has its two ports in line with each other on the opposite sides of the valve, whereas the seat-type has the in/out ports located at a 90 degree angle from one another.

The saddle type is the most common in process applications and the seat-type is more commonly used as a tank bottom valve but exists also as a process valve.

While diaphragm valves usually come in two-port forms (2/2-way diaphragm valve), they can also come with three ports (3/2-way diaphragm valves also called t-valves) and more (so called block-valves). When more than three ports are included, they generally require more than one diaphragm seat; however, special dual actuators can handle more ports with one membrane.

Diaphragm valves can be manual or automated. Their application is generally as shut-off valves in process systems within the food and beverage, pharmaceutical and biotech industries. The older generation of these valves is not suited for regulating and controlling process flows, however newer developments in this area have successfully tackled this problem.

In addition to the well known, two way shut off diaphragm valve, other types include: three way zero deadleg valve, sterile access port, block and bleed, valbow and tank bottom valve just to name a few.

Actuators

Diaphragm valves can be controlled by various types of actuators e.g. manual, pneumatic, hydraulic, electric etc. The most common diaphragm valves use pneumatic actuators; in this type of valve, air pressure is applied through a Schrader valve which raises the diaphragm and opens the valve. This type of valve is extremely quick and as such is one of the more common valves used in operations where valve speed is a necessity.

Hydraulic diaphragm valves also exist for higher pressure and lower speed operations. Some diaphragm valves are also controlled manually.

Body Materials

- Brass
- Steel type:
  - Cast Iron
  - Ductile Iron
  - Carbon Steel
  - Stainless Steel.
- Plastic type:
  - ABS (Acrylonitrile butadiene styrene)

    - o PVC-U (Polyvinyl chloride, unplasticized) also known as PVCu or uPVC
    - o PVC-C (Polyvinyl chloride, post chlorinated) also known as PVCc or cPVC
    - o PP (Polypropylene)
    - o PE (Polyethylene) also known as LDPE, MDPE and HDPE
    - o PVDF (Polyvinylidene fluoride).

Body Lining Materials

Depending on temperature, pressure and chemical resistance, one of the following is used:

- Unlined type
- Rubber lined type:
    - o NR/Hard Rubber/Ebonite,
    - o BR/Soft rubber
    - o EPDM
- Fluorine plastic lined type
- o FEP/F46
- o PFA
- o PO.

Diaphragm Materials

- Unlined or Rubber Lined Type:
    - o NR/Natural Rubber
    - o NBR/Nitrile/Buna-N
    - o EPDM
    - o FKM/Viton
    - o SI/Silicone rubber.
- Fluorine Plastic Type:
    - o FEP/F46,with EPDM back
    - o PTFE/F4,with EPDM back
    - o PFA,with EPDM back.

Gate Valve

A gate valve, also known as a sluice valve, is a valve that opens by lifting a round or rectangular gate/wedge out of the path of the fluid. The distinct feature of a gate valve is the sealing surfaces between the gate and seats are planar, so gate valves are often used when a straight-line flow of fluid and minimum restric-tion is desired. The gate faces can form a wedge shape or they can be parallel. Gate valves are primarily used to permit or prevent the flow of liquids, but typical gate valves shouldn't be used for regulating flow, unless they are specifically designed for that purpose. Because of their ability to cut through liquids, gate valves are often used in the petroleum industry. On opening the gate valve, the flow path is enlarged in a highly nonlinear manner with respect to percent of opening.

This means that flow rate does not change evenly with stem travel. Also, a partially open gate disk tends to vibrate from the fluid flow. Most of the flow change occurs near shutoff with a relatively high fluid velocity causing disk and seat wear and eventual leakage if used to regulate flow. Typical gate valves are designed to be fully opened or closed. When fully open, the typical gate valve has no obstruction in the flow path, resulting in very low friction loss.

Figure: *A stainless steel gate valve*

Gate valves are characterised as having either a rising or a nonrising stem. Rising stems provide a visual indication of valve position because the stem is attached to the gate such that the gate and stem rise and lower together as the valve is operated. Nonrising stem valves may have a pointer threaded onto the upper end of the stem to indicate valve position, since the gate travels up or down the stem on the threads without raising or lowering the stem. Nonrising stems are used underground or where vertical space is limited.

Bonnets provide leakproof closure for the valve body. Gate valves may have a screw-in, union, or bolted bonnet. Screw-in bonnet is the simplest, offering a durable, pressure-tight seal. Union bonnet is suitable for applications requiring frequent inspection and cleaning. It also gives the body added strength. Bolted bonnet is used for larger valves and higher pressure applications.

Another type of bonnet construction in a gate valve is pressure seal bonnet. This construction is adopted for valves for high pressure service, typically in excess of 15 MPa (2250 psi). The unique feature about the pressure seal bonnet is that the body-bonnet joints seals improves as the internal pressure in the valve increases, compared to other constructions where the increase in internal pressure tends to create leaks in the body-bonnet joint. Gate valves may have flanged ends which are drilled according to pipeline compatible flange dimensional standards. Gate valves are typically constructed from cast iron, ductile iron, cast carbon steel, gun metal, stainless steel, alloy steels, and forged steels.

Globe Valve

A globe valve is a type of valve used for regulating flow in a pipeline, consisting of a movable disk-type element and a stationary ring seat in a generally spherical body.

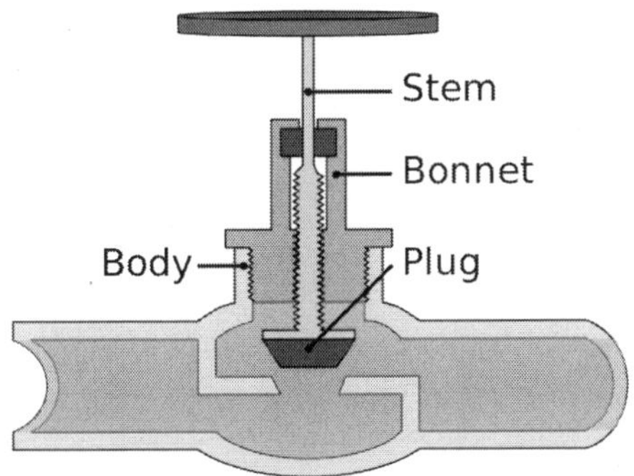

Figure: *Internal parts of a typical globe valve.*

Globe valves are named for their spherical body shape with the two halves of the *body* being separated by an internal baffle. This has an opening that forms a *seat* onto which a movable *plug* can be screwed in to close (or shut) the valve. The plug is also called a *disc* or *disk*. In globe valves, the plug is connected to a *stem* which is operated by screw action in manual valves. Typically, automated valves use sliding stems. Automated globe valves have a smooth stem rather than threaded and are opened and closed by an actuator assembly. When a globe valve is manually operated, the stem is turned by a handwheel.

Although globe valves in the past had the spherical bodies which gave them their name, many modern globe valves do not have much of a spherical shape. However, the term *globe valve* is still often used for valves that have such an internal mechanism. In plumbing, valves with such a mechanism are also often called stop valves since they don't have the global appearance, but the term *stop valve* may refer to valves which are used to stop flow even when they have other mechanisms or designs.

Globe valves are used for applications requiring throttling and frequent operation. For example, globe valves or valves with a similar mechanism may be used as sampling valves, which are normally shut except when liquid samples are being taken. Since the baffle restricts flow, they're not recommended where full, unobstructed flow is required.

Parts of a Typical Globe Valve

Body: The body is the main pressure containing structure of the valve and the most easily identified as it forms the mass of the valve. It contains all of the valve's internal parts that will come in contact with the substance being controlled by the valve. The bonnet is connected to the body and provides the containment of the fluid, gas, or slurry that is being controlled.

Globe valves are typically two-port valves, although three port valves are also produced mostly in straight-flow configuration. *Ports* are openings in the body for fluid flowing in or out. The two ports may be oriented straight across from each other on the body, or oriented at an angle such as a 90° angle. Globe valves with ports at such an angle are called angle globe valves. Angle type globe valves are mainly used for corrosive or high viscous fluid which shall solidify at room temperature. This is because straight valve shall have outlet pipe inline to the inlet pipe and the fluid has chance of staying there in case of horizontal piping whereas in the case of angle valve the outlet pipe shall be directed towards bottom which will allow the fluid to drain of instead of staying there which will in turn clog or corrode the valve components when allowed to settle down. A globe valve can also have a body in the shape of a "Y". This will allow the construction of valve to be straight at the bottom instead of conventional pot type construction (to arrange bottom seat) in case of other valves. This will again allow the fluid to pass through without low cavities which shall let the fluid clog or corrode in long run of settling down.

Bonnet: The bonnet provides a leakproof closure for the valve body. The threaded section of the stem goes through a hole with matching threads in the bonnet. Globe valves may have a screw-in, union, or bolted bonnet. Screw-in bonnet is the simplest bonnet, offering a durable, pressure-tight seal. Union bonnet is suitable for applications requiring frequent inspection or cleaning. It also gives the body added strength. A bonnet attached with bolts is used for larger or higher pressure applications. The bonnet also contains the packing, a wearable material that maintains the seal between the bonnet and the stem during valve cycling.

Plug or Disc (Disk): The closure member of the valve, plugs are connected to the stem which is slid or screwed up or down to throttle the flow. Plugs are typically of the balance or unbalanced type. Unbalanced plugs are solid and are used with smaller valves or with low pressure drops across the valve. The advantages are simpler design, with one possible leak path at the seat and usually lower cost. The disadvantages are the limited size; with a large unbalanced plug the forces needed to seat and hold the flow often becomes impractical. Balanced plugs have holes through the plug. Advantages include easier shut off as the plug does not have to overcome static forces. However, a second leak path is created between the plug and the cage, and cost is generally higher.

Stem: The stem serves as a connector from the actuator to the inside of the valve and transmits this actuation force. Stems are either smooth for actuator controlled valves or threaded for manual valves. The smooth stems are surrounded by packing material to prevent leaking material from the valve. This packing is a wearable material and will have to be replaced during maintenance. With a smooth stem the ends are threaded to allow connection to the plug and the actuator. The stem must not only withstand a large amount of compression force during valve closure, but also have high tensile strength during valve opening. In addition, the stem must be very straight, or have low run out, in order to ensure good valve closure. This minimum run out also minimizes wear of the packing containedty in the bonnet, which provides the seal against leakage. The stem may be provided with a shroud over the packing nut to prevent foreign bodies entering the packing material, which would accelerate wear.

Cage: The cage is part of the valve that surrounds the plug and is located inside the body of the valve. Typically, the cage is one of the greatest determiners of flow within the valve. As the plug is moved more of the openings in the cage are exposed and flow is increased and vice versa. The design and layout of the openings can have a large effect on flow of material (the flow characteristics of different materials at temperatures, pressures that are in a range). Cages are also used to guide the plug to the seat of the valve for a good shutoff, substituting the guiding from the bonnet.

Seat Ring: The seat ring provides a stable, uniform and replaceable shut off surface. Seat rings are usually held in place by pressure from the fastening of the bonnet to the top of the body. This pushes the cage down on the lip of the seat ring and holds it firmly to the body of the valve. Seat rings may also be threaded and screwed into a thread cut in the same area of the body. However this method makes removal of the seat ring during maintenance difficult if not impossible. Seat rings are also typically beveled at the seating surface to allow for some guiding during the final stages of closing the valve.

Economical globe valves or stop valves with a similar mechanism used in plumbing often have a rubber washer at the bottom of the disc for the seating surface, so that rubber can be compressed against the seat to form a leak-tight seal when shut.

Materials

Typically globe valves are made of metallic alloys, although some synthetic materials are available. These materials are chosen based on pressure, temperature, controlled media properties. Corrosive and/or erosive process streams may require a compromise in material selection or exotic alloys or body coatings to minimize these material interactions and extend the life of the valve or valve trim components. Typically, carbon steel alloys are specified for noncorrosive applications. Other alloys such as Hastelloy, Monel, Inconel and others are available.

Packing material must also be considered during valve selection. Typically the requirement for a low friction packing conflict with a durable material that will provide low maintenance requirements during service life. Corrosive applications can further complicate packing material selection as the typical packing materials may or may not be compatible with the processed materials. Typically graphite or PTFE is used due to its low friction coefficient. Enviro-seal applications also have the availability of constant applied force (live-load) packing. While more complex, it allows for constant packing force load throughout the life of the packing material. This packing helps meet contemporary environmental laws.

Needle Valve

A needle valve is a type of valve having a small port and a threaded, needle-shaped plunger. It allows precise regulation of flow, although it is generally only capable of relatively low flow rates.

Construction and Operation

A needle valve has a relatively small orifice with a long, tapered seat, and a needle-shaped plunger, on the end of a screw, which exactly fits this seat. As the screw is turned and the plunger retracted, flow between the seat and the plunger is possible; however, until the plunger is completely retracted the fluid flow is significantly impeded. Since it takes many turns of the fine-threaded screw to retract the plunger, precise regulation of the flow rate is possible.

The virtue of the needle valve is from the vernier effect of the ratio between the needle's length and its diameter, or the difference in diameter between needle and seat. A long travel axially (the control input) makes for a very small and precise change radially (affecting the resultant flow). Needle valves may be used in vacuum systems, when a very precise control of gas flow is required, at low pressure, such as when filling gas-filled vacuum tubes, gas lasers and similar devices.

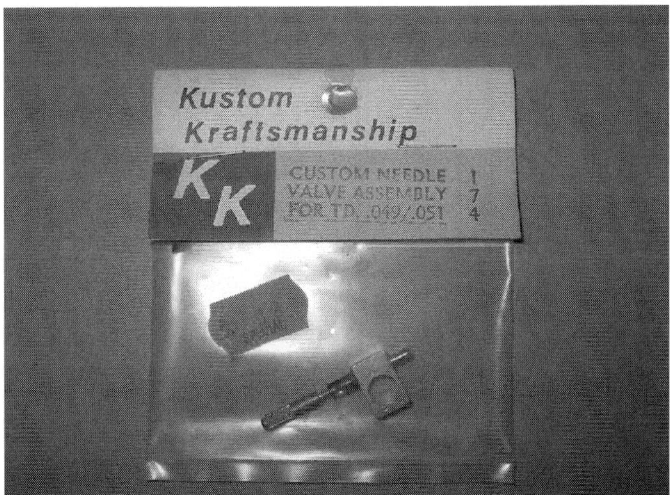

Figure: *A needle valve.*

Uses

Needle valves are usually used in flow metering applications, especially when a constant, calibrated, low flow rate must be maintained for some time, such as the idle fuel flow in a carburetor. Note that the float valve of a carburetor (controlling the fuel level within the carburetor) is *not* a needle valve, although it is commonly described as one. It uses a bluntly conical needle, but it seats against a square-edged seat rather than a matching cone. The intention here is to obtain a well-defined seat between two narrow mating surfaces, giving firm shutoff of the flow from only a light float pressure.

Since flow rates are low and many turns of the valve stem are required to completely open or close, needle valves are not used for simple shutoff applications.

Since the orifice is small and the force advantage of the fine-threaded stem is high, needle valves are usually easy to shut off completely, with merely "finger tight" pressure. Small, simple needle valves are often used as bleed valves in hot water heating applications. Unlike a ball valve, or valves with a rising stem, it is not easy to tell from examining the handle position whether the valve is open or closed.

Pinch Valve

A pinch valve is a full bore or fully ported type of control valve which uses a pinching effect to obstruct fluid flow. There are a few types of pinch valves based upon application.

Fluids

Pinch valves used for fluids usually employ a device that directly contacts process tubing. Forcing the tubing together will create a seal that is equivalent to the tubing's permeability.

Particulates

Major components of a pinch valve consists of body and a sleeve. The sleeve will contain the flow media and isolate it from the environment hence reducing contamination to the environment. Generally used for slurries or processes with entrained solids, because the flexible rubber sleeve allows the valve to close droptight around solids—solids that would typically be trapped by the seat or stuck in crevices in globe, diaphragm, butterfly, gate, or ball valves. The sleeve material can be selected upon the corrosiveness and abrasiveness of the flow media, a suitable synthetic polymer can be chosen. A pinch valve may be the best type of valve for flow control application if the operation temperature is within the limit of the polymer.

Air Operated Pinch Valves

Air operated Pinch Valves are very easily constructed. They generally only consist of an extremely elasticised reinforced rubber hose, a type of housing, and two socket end covers or flanges. The rubber hoses of the air operated Pinch Valves are usually press-fitted and centred into the housing ends by the socket end covers or flanges.

Function of Air Operated Pinch Valves

The air operated Pinch Valve works without any additional actuator; all it needs to close or operate is 30 psi air supply into the

Pinch Valve body. As soon as the air supply becomes interrupted and the volume of air exhausts, the elastic rubber hose starts to open due to its great impact resilience.

Advantage of Air Operated Pinch Valves

The most important benefits of using air operated Pinch Valves are their complete and true full bore, and the 100% tight shut off – even on solids such as granules, powders, pellets, chippings, fibres, slivers, any kind of slurries and many more aggressive products.

Usually conventional valves such as ball valves, butterfly valves, piston valves or gate valves tend to fail when in process with the above mentioned aggressive products. The reason is that the valve body seat or the gate/piston of the valves wears out too quickly and the valve can no longer shut off tightly, whereas air operated Pinch Valves work without almost any wear of the elastic rubber hose because the kinetic energy of the solids are absorbed through the extremely high elasticity of the rubber. The rubber mixture is also anti abrasive.

Industrial Sectors

Therefore using air operated Pinch Valves in aggressive product processes saves a lot down time, and guarantees a free flow through the valve, avoiding any blockages.

Several rubber qualities are available for air operated Pinch Valves such as natural rubber, EPDM, nitrile, viton, neoprene and butyl. Different housings and end covers/flange materials such as aluminium, plastics and stainless steel are also available; resulting in air operated Pinch Valves being commonly used in a wide range of many different industries such as:

- Bulk and solids handling,
- Cement industry,
- Waste water industry,
- Chemical industry,
- Food industry,
- Beverage industry,
- Ceramic-/Glass-/Plastic industry.

Piston Valve

A ‘piston valve’ is a device used to control the motion of a fluid along a tube or pipe by means of the linear motion of a piston within a chamber or cylinder.

Examples of piston valves are:

- The valves used in the valve gear of many stationary steam engines and most steam locomotives.
- The valves used in many brass instruments.
- The valves used in pneumatic cannons.

Steam Engines

The Swannington incline winding engine on the Leicester and Swannington Railway, manufactured by The Horsely Coal & Iron Company in 1833, shows a very early use of the piston valve. Piston valves had been used a year or two previously in the horizontal engines manufactured by Taylor and Martineau of London, but did not become general for stationary or locomotive engines until the end of the 19th century. In the 19th century, most steam locomotives used slide valves to control the flow of steam into and out of the cylinders. In the 20th century, slide valves were gradually superseded by piston valves, particularly in engines using superheated steam. There were two reasons for this:

- With piston valves, the steam passages can be made shorter. This reduces resistance to the flow of steam and improves efficiency
- It is difficult to lubricate slide valves adequately in the presence of superheated steam

The usual locomotive valve gears, *e.g.*, the Stephenson valve gear, Walschaerts valve gear, and Baker valve gear, can be used with either slide valves or piston valves. Where poppet valves are used, a different gear, such as Caprotti valve gear is needed.

Brass Instruments

Cylindrical piston valves called Perinet valves (after their inventor, Franηois Perinet), are used to change the length of tube in the playing of most brass instruments, particularly the trumpet-like members of the family (cornet, flugelhorn, saxhorn, etc.).

Other brass instruments use rotary valves. notably the orchestral horns and many tuba models, but also a number of rotary-valved variants of those brass instruments which more commonly employ piston valves.

The first piston-valved musical instruments were developed just after the start of the 19th century. The Stölzel valve (invented by Heinrich Stölzel in 1814) was an early variety. In the mid 19th century the Vienna valve was an improved design. However most professional musicians preferred rotary valves for quicker, more reliable action, until better designs of piston valves were mass manufactured in towards the end of the 19th century.

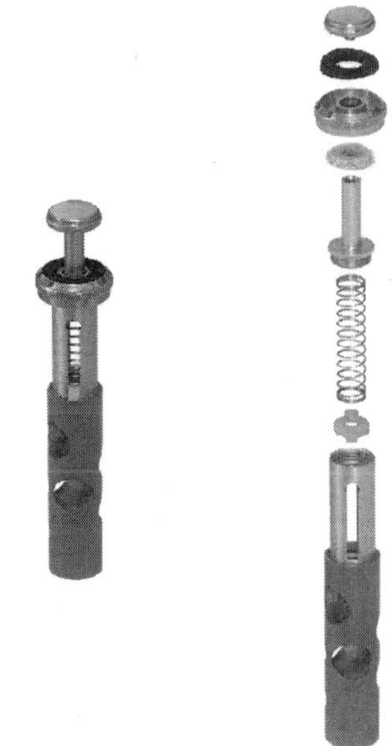

Figure: *A trumpet's piston valve*

Pneumatic Cannons

A piston valve can also refer to a 2-way 2-position, pilot-operated spool valve. The term is extremely popular among spud gun enthusiasts who often build homemade piston valves for use in pneumatic cannons. Valves are typically constructed primarily from pipe fittings and machined plastics or metals. The inside of a piston valve contains a piston that blocks the output when the valve is pressurized, and a volume of air behind the piston. When the pressure behind the piston is released the piston is pushed back by the force of the pressure from the input. This allows the valve to be opened by a much smaller pilot valve, with speeds faster than possible with just a manually operated valve. Functionally these types of valves are comparable to quick exhaust valves.

This type of piston valve is also sometimes referred to as a back-pressure valve.

Plug Valve

Plug valves are valves with cylindrical or conically-tapered "plugs" which can be rotated inside the valve body to control flow through the valve. The plugs in plug valves have one or more hollow passageways going sideways through the plug, so that fluid can flow through the plug when the valve is open. Plug valves are simple and often economical.

When the plug is conically-tapered, the stem/handle is typically attached to the larger diameter end of the plug. Plug valves usually do not have bonnets but often have the end of the plug with the handle exposed or mostly exposed to the outside. In cases like that, there is usually not much of a stem. The stem and handle often come in one piece, often a simple, approximately L-shaped handle attached to the end of the plug. The other end of the plug is often exposed to the outside of the valve too, but with a mechanism which retains the plug in the body.

The simplest and most common general type of plug valve is a 2-port valve, which has two positions, *open* to allow flow, and *shut* (closed) to stop flow. Ports are openings in the valve body through which fluid can enter or leave. The plug in this kind of valve has one passageway going through it. The ports are typically at opposite ends of the body; therefore, the plug is rotated a fourth of a full turn to change from open to shut positions. This makes this kind of plug valve a *quarter-turn valve*. There is often a mechanism limiting motion of the handle to a quarter turn, but not in glass stopcocks.

Slightly conically-tapered metal (often brass) plug valves are often used as simple shut-off valves in household natural gas lines.

It is also possible for a plug valve to have more than two ports. In a 3-way plug valve, flow from one port could be directed to either the second or third port. A 3-way plug valve could also be designed to shift flow between ports 1 and 2, 2 and 3, or 1 and 3, and possibly even connect all three ports together. The flow-directing possibilities in multi-port plug valves are similar to the possibilities in corresponding multi-port ball valves or corresponding multi-port valves with a rotor. An additional possibility in plug valves is the have one port on one side of the plug valve and two ports on the other side, with two diagonal and parallel fluid pathways inside the plug. In this case the plug can be rotated 180° to connect the port on the one side to either of the two ports on the other side.

Stopcocks used in laboratory glassware are typically forms of conically-tapered plug valves. When fused with the glassware, the valve bodies are made of glass. Otherwise, they can be made of an inert plastic such as Teflon. The plugs can be made of a similar plastic or glass. When the plug is made of glass, the handle and plug are fused together in one piece out of glass. When glass is used for both the stopcock body and the plug, the contacting surfaces between them are special ground glass surfaces often with stopcock grease in between. Special glass stopcocks are made for vacuum applications, such as in use with vacuum manifolds. Stopcock grease is always used in high vacuum applications to make the stopcock air-tight. Also if the plug valve is "locked" from being in the open or closed position for an extended amount of time lubricant can be added through the greaser with the valve in service.

Poppet Valve

A poppet valve (also called mushroom valve) is a valve consisting of a hole, usually round or oval, and a tapered plug, usually a disk shape on the end of a shaft also called a valve stem. The shaft guides the plug portion by sliding through a valve guide. In most applications a pressure differential helps to seal the valve and in some applications also open it.

Presta and Schrader valves used on pneumatic tires are examples of poppet valves. The Presta valve has no spring and relies on a pressure differential for opening and closing while being inflated.

Etymology

The word poppet shares etymology with "puppet": it is from the Middle English *popet* ("youth" or "doll"), from Middle French *poupette*, which is a diminutive of *poupe*. The use of the word *poppet* to describe a valve comes from the same word applied to marionettes, which – like the poppet valve – move bodily in response to remote motion transmitted linearly. In the past, "puppet valve" was a synonym for *poppet valve*; however, this usage of "puppet" is now obsolete.

Operation

The operating principle of poppet valves is described in "How Poppet Valves Work". In most cases it is beneficial to have a "balanced poppet" in a direct-acting valve. Less force is needed to move the poppet because all forces on the poppet are nullified by equal and opposite forces. The solenoid coil has to counteract only the spring force.

Applications

Poppet valves are used in many industrial processes, from controlling the flow of milk to isolating sterile air in the semiconductor industry. However, they are most well known for their use in internal combustion and steam engines.

Internal Combustion Engine

Poppet valves are used in most piston engines to open and close the intake and exhaust ports in the cylinder head. The valve is usually a flat disk of metal with a long rod known as the *valve stem* attached to one side. The stem is used to push down on the valve and open it, with a spring generally used to return it to the closed position when the stem is not being depressed. At high RPM, spring forces are too slow to return the valve to its seat between cycles, leading to 'valve float'. In this situation desmodromic valves are used which, being closed by a positive mechanical action instead of by a spring, are able to cycle at the high speeds required in, for instance, motorcycle and auto racing engines.

The engine normally operates the valves by pushing on the stems with cams and cam followers. The shape and position of the cam determines the valve lift and when and how quickly (or slowly) the valve is opened. The cams are normally placed on a fixed camshaft which is then geared to the crankshaft, running at half crankshaft speed in a four-stroke engine. On high-performance engines, the camshaft is movable and the cams have a varying height, so by axially moving the camshaft in relation with the engine RPM, also the valve lift varies.

For certain applications the valve stem and disk are made of different steel alloys, or the valve stems may be hollow and filled with sodium to improve heat transport and transfer. Although better heat conductors, aluminium cylinder heads require steel valve seat inserts, while cast-iron cylinder heads often used integral valve seats in the past. Because the valve stem extends into lubrication in the cam chamber, it must be sealed against blow-by to prevent cylinder gases from escaping into the crankcase. A rubber lip-type seal ensures that excessive amounts of oil are not drawn in from the crankcase on the induction stroke and that exhaust gas does not enter the crankcase on the exhaust stroke. Worn valve seals are characterised by a puff of blue smoke from the exhaust when pressing back down on the accelerator pedal after allowing the engine to over-run, such as when changing gears.

Valve Position

In very early engine designs the valves were 'upside down' in the block, parallel to the cylinders-the so called L-head engine because of the shape of the cylinder and combustion chamber, also called 'flathead engine' as the top of the cylinder head is flat. Although this design makes for simplified and cheap construction, it has two major drawbacks; the tortuous path followed by the intake charge limits air flow and effectively prevents speeds greater than 2,000–2,500 RPM, and the travels of the exhaust through the block can cause overheating under sustained heavy load. This design evolved into 'Intake Over

Exhaust', IOE or F-head, where the intake valve was in the head and the exhaust valve was in the block; later both valves moved to the head. In most such designs the camshaft remained relatively near the crankshaft, and the valves were operated through pushrods and rocker arms. This led to significant energy losses in the engine, but was simpler, especially in a V engine where one camshaft can actuate the valves for both cylinder banks; for this reason, pushrod engine designs persisted longer in these configurations than others.

More modern designs have the camshaft on top of the cylinder head, pushing directly on the valve stem (again through cam followers, also known as tappets), a system known as *overhead camshaft*; if there is just one camshaft, this is a single overhead cam or *SOHC* engine. Often there are two camshafts, one for the intake and one for exhaust valves, creating the dual overhead cam, or *DOHC*. The camshaft is driven by the crankshaft-through gears, a chain or a timing belt.

Valve Wear

In the early days of engine building, the poppet valve was a major problem. Metallurgy was not what it is today, and the rapid opening and closing of the valves against the cylinder heads led to rapid wear. They would need to be re-ground every two years or so, in an expensive and time-consuming process known as a *valve job*. Adding tetra-ethyl lead to the petrol reduced this problem to some degree, as the lead would coat the valve seats, in effect lubricating the metal. Valve seats made of improved alloys such as stellite have generally made this problem disappear completely and made leaded fuel unnecessary.

Steam Engine

When used in high-pressure applications, for example, as admission valves on steam engines, the same pressure that helps seal poppet valves also contributes significantly to the force required to open them. This has led to the development of the balanced poppet or double beat valve, in which two valve plugs ride on a common stem, with the pressure on one plug largely balancing the pressure on the other. In these valves, the force needed to open the valve is determined by the pressure and the difference between the areas of the two valve openings.

Poppet valves have been used on steam locomotives, often in conjunction with Lentz or Caprotti valve gear. British examples include:

- LNER Class B12
- LNER Class D49
- LNER Class P2

- LMS Stanier Class 5 4-6-0
- BR standard class 5
- BR standard class 8 71000 Duke of Gloucester.

Sentinel Waggon Works used poppet valves in their steam wagons and steam locomotives. Reversing was achieved by a simple sliding camshaft system.

The poppet valve was also used on the American Pennsylvania Railroad's T1 duplex locomotives, although the valves commonly failed due to the fact that the locomotives were commonly operated in excess of 160 km/h (100 mph), and the valves were not meant for the stresses of such speeds.

Safety Valve

A safety valve is a valve mechanism for the automatic release of a substance from a boiler, pressure vessel, or other system when the pressure or temperature exceeds preset limits.

It is part of a bigger set named pressure safety valves (PSV) or pressure relief valves (PRV). The other parts of the set are named relief valves, safety relief valves, pilot-operated relief valves, low pressure safety valves, vacuum pressure safety valves.

Safety valves were first used on steam boilers during the industrial revolution. Early boilers without them were prone to accidental explosion.

Vacuum safety valves (or combined pressure/vacuum safety valves) are used to prevent a tank to collapse when emptying it or when cold rinse water is used after hot CIP or SIP. The calculation method is not defined in any norm when sizing a vacuum safety valve, particularly in the hot CIP/cold water scenario, but some manufacturers have developed simulations to do so.

Function and Design

The earliest and simplest safety valve on the steam digester in 1679 used a weight to hold the pressure of the steam, (this design is still commonly used on pressure cookers); however, these were easily tampered with or accidentally released. On the Stockton and Darlington Railway, the safety valve tended to go off when the engine hit a bump in the track. A valve less sensitive to sudden accelerations used a spring to contain the steam pressure, but these (based on Salter spring balances) could still be screwed down to increase the pressure beyond design limits. This dangerous practice was sometimes used to marginally increase performance of a steam engine. In 1856 John Ramsbottom invented a tamper-proof spring safety valve which became universal on railways.

Safety valves also evolved to protect equipment such as pressure vessels (fired or not) and heat exchangers. The term safety valve should be limited to compressible fluid application (gas, vapour, steam).

The two general types of protection encountered in industry are *thermal protection* and *flow protection*.

For liquid-packed vessels, thermal relief valves are generally characterized by the relatively small size of the valve necessary to provide protection from excess pressure caused by thermal expansion. In this case a small valve is adequate because most liquids are nearly incompressible, and so a relatively small amount of fluid discharged through the relief valve will produce a substantial reduction in pressure. Flow protection is characterized by safety valves that are considerably larger than those mounted in thermal protection. They are generally sized for use in situations where significant quantities of gas or high volumes of liquid must be quickly discharged in order to protect the integrity of the vessel or pipeline. This protection can alternatively be achieved by installing a high integrity pressure protection system (HIPPS).

Technical Terms

In the petroleum refining, petrochemical, chemical manufacturing, natural gas processing, power generation, food, drinks, cosmetics and pharmaceuticals industries, the term safety valve is associated with the terms pressure relief valve (PRV), pressure safety valve (PSV) and relief valve. The generic term is Pressure relief valve (PRV) or pressure safety valve (PSV) It should be noted, as most people think PRV and PSV are the same thing, they are not. The difference is that PSV's have a manual lever to open the valve in case of emergency.

- Relief valve (RV): automatic system that is actuated by static pressure in a liquid-filled vessel. It specifically opens proportionally with increasing pressure.
- Safety valve (SV): automatic system that relieves the static pressure on a gas. It usually opens completely, accompanied by a popping sound.
- Safety relief valve (SRV): automatic system that relieves by static pressure on both gas and liquid.
- Pilot-operated safety relief valve (POSRV): automatic system that relieves by remote command from a pilot on which the static pressure (from equipment to protect) is connected.

- Low pressure safety valve (LPSV): automatic system that relieves static pressure on a gas. Used when difference between vessel pressure and the ambient atmospheric pressure is small.
- Vacuum pressure safety valve (VPSV): automatic system that relieves static pressure on a gas. Used when the pressure difference between the vessel pressure and the ambient pressure is small, negative and near the atmospheric pressure.
- Low and vacuum pressure safety valve (LVPSV): automatic system that relieves static pressure on a gas. The pressure is small, negative or positive and near the atmospheric pressure.

RV, SV and SRV are spring operated (even said spring loaded). LPSV and VPSV are spring operated or weight loaded.

Legal and Code Requirements in Industry

In most countries, industries are legally required to protect pressure vessels and other equipment by using relief valves. Also in most countries, equipment design codes such as those provided by the ASME, API and other organizations like ISO (ISO 4126) must be complied with and those codes include design standards for relief valves.

Today, industries such as food, drinks, cosmetics, pharmaceuticals and fine chemicals industries ask for hygienic safety valves, fully drainable and Cleanable In Place; most of them are stainless steel made; Hygienic norms are mainly 3A in the US and EHEDG in Europe.

Types

There is a wide range of safety valves having many different applications and performance criteria required to cover different areas. In addition, national standards are set by many kinds of safety valve.

- ASME I tap-a safety valve in accordance with the requirements of Section I of the application code ASME pressure vessel, which opens 3% and 4% of the pressure. Will rule on two rings serving and supported by a National Seal V defined.
- ASME VIII valve-safety valve in accordance with the requirements of Article VIII of the ASME code for pressure vessel applications, which is within 10% overpressure that opens and closes in 7%. Characterized by a National Board UV stamp.
- Low-lift safety valve-the current position of the disc around the drain valve.
- Full lift safety valve-the region of the exemption is not determined by the position of the disc.

- Full-flow relief valve-A valve which is expected in the hole and lift the valve a sufficient measure of the minimum area for each position or under the seat to make the control panel.

  Safety Valve * classic-The spring housing is vented to the pressure side, i.e., functional characteristics are directly influenced by changes in pressure in the valve.
- Valve balanced-a balanced valve includes a means to minimize the effects of pressure on the operating characteristics of the valve.
- Guide pressure valve-The largest dump device is in conjunction with controlled, self-actuated auxiliary pressure relief.
- Power operated valve-A pressure relief valve in which the main system to relieve pressure combined with and controlled by a device requires an external power source.
- Is the standard safety valve-a valve opening to achieve the degree of buoyancy required for mass flow increases the pressure to reject over 10%. (The valve is characterized by a score of popular art, sometimes also called high-lift).
- Full-lift (solid line) valve-a valve that opens at the beginning of the elevator, quickly, with a 5% increase in pressure until complete removal is limited by design. The amount of lift to an early start (analog) can not exceed 20%.
- Valve and spring-a safety valve for the treatment of violent opening of the valve plate by a clamping force as a spring or weight.
- Proportional-relief valve-a safety valve which opens more or less stable in comparison with increasing pressure. sudden opening within a stroke by 10% will not happen without increasing pressure. After opening at a pressure greater than 10% meet these safety valves of the lift and landed on the mass flowmeter.
- Safety valve diaphragm-A direct-loaded safety valve with linear and rotary components and springs to protect against the effects of the liquid membrane.
- Bellows valve-A direct-loaded safety valve and slide (or part) of rotating elements and sources are protected from the effects of fluid through bellows. The bellows allows such an interpretation, it is as if the influences of pressure.
- Controlled valve-A valve that consists of a main valve and a control unit. It also includes direct acting safety valves with

additional load, which, until the total pressure reached, an additional force increases the closing force.

- Safety valve-A valve that automatically, without using any form of energy than the liquid in the discharge quantity of liquid, so as to prevent a predetermined safe pressure is higher, and closed again to prevent the flow of more volatile after the normal working pressure were restored. Note that the valve can be characterized by a pop-action (quick opening) or open relationship (not necessarily linear) on increasing pressure over the whole pressure.
- Valve and spring-a safety valve when the charger through the liquid under pressure valve plate with mechanical pressure directly, unlike, for example, a weight, lever and weight or spring. * Assisted valve-A valve which can be lifted by a support mechanism for moving, also gradually the pressure that the pressure setting, and for failure to comply with a mechanism to support all requirements for safety valves in the standard.
- Safety valve spring in addition-as a safety valve until the inlet pressure safety valve to the pressure achieved additional power increases the power of foreclosure. Note that this extra power (extra weight), which is made from an external power source is available, reliable published when the inlet pressure reaches the safety valve pressure set. The amount of the extra load is placed so that if such a charge is released, the safety valve of certified capacity at a pressure not greater than 1. protected 1 times the maximum allowable pressure devices.
- Master valve-A valve whose operation is initiated and controlled by the fluid discharged from a pilot valve is a safety net direct costs, provided that standard.

United States

- ASME (American Society of Mechanical Engineers) Boiler & Pressure Vessel Code, Section I
- ASME (American Society of Mechanical Engineers) Boiler & Pressure Vessel Code, Section VIII, Division 1
- API (American Petroleum Institute) Recommended Practice 520 and API Standard 526, API Standard 2000 (low pressure-Storage tank).

European Union

- ISO 4126 (harmonized with European Union directives)

- EN 764-7 (former CEN standard, harmonized with European Union directives, replaced with EN ISO 4126-1)
- AD Merkblatt (German)
- PED 97/23/EC (Pressure Equipment Directive-European Union).

Japan

- JIS: Japanese Industrial Standards.

South Korea

- KOSHA.

Water Heaters

They are required on water heaters, where they prevent disaster in certain configurations in the event a thermostat should fail. There are still occasional, spectacular failures of older water heaters that lack this equipment. Houses can be leveled by the force of the blast.

Pressure Cookers

Pressure cookers are pots for cooking with a pressure-proof lid. Cooking at pressure allows the temperature to rise above the normal boiling point of water (100 degrees Celsius at sea level) which speeds up the cooking and makes the cooking more thorough.

Pressure cookers usually have two safety valves. One is a hole upon which a weight sits. The other is a sealed rubber grommet which is ejected in a controlled explosion if the first valve gets blocked.

Specific Types

- Aspin valve: a cone-shaped metal part fitted to the cylinder head of an engine
- Ball cock: often used as a water level controller (cistern)
- Bibcock: provides a connection to a flexible hosepipe
- Blast valve: prevents rapid overpressures in a fallout shelter or a bunker
- Cock: colloquial term for a small valve or a stopcock
- Demand valve: on a diving regulator
- Double beat valve
- Double check valve
- Duckbill valve
- Flipper valve

- Flow control valve: an application which maintains a variable flow rate through the valve
- Heimlich valve: a specific one-way valve used on the end of chest drain tubes to treat a pneumothorax
- Foot valve: a check valve on the foot of a suction line to prevent backflow
- Four-way valve: was used to control the flow of steam to the cylinder of early double-acting steam engines
- Freeze seal/Freeze plug: in which freezing and melting the fluid creates and removes a plug of frozen material acting as the valve
- Gas pressure regulator regulates the flow and pressure of a gas
- Heart valve: regulates blood flow through the heart in many organisms
- Leaf valve: one-way valve consisting of a diagonal obstruction with an opening covered by a hinged flap
- Pilot valve: regulate flow or pressure to other valves
- Plunger Valve: To regulate flow while lowering the pressure
- Poppet valve and sleeve valve: commonly used in piston engines to regulate the fuel mixture intake and exhaust
- Pressure regulator or pressure reducing valve (PRV): reduces pressure to a preset level downstream of the valve
- Pressure sustaining valve, or back-pressure regulator: maintains pressure at a preset level upstream of the valve
- Presta and Schrader valves are used to hold the air in bicycle tires
- Reed valve: consists of two or more flexible materials pressed together along much of their length, but with the influx area open to allow one-way flow, much like a heart valve
- Regulator: used in SCUBA diving equipment and in gas cooking equipment to reduce the high pressure gas supply to a lower working pressure
- Rocker valve
- Rotary valves and piston valves: parts of brass instruments used to change their pitch
- Rupture disc: a one time use replaceable valve for rapid pressure relief, used to protect piping systems from excessive pressure or vacuum; more reliable than a safety valve

- Saddle valve: where allowed, is used to tap a pipe for a low-flow need
- Safety valve or relief valve: operates automatically at a set differential pressure to correct a potentially dangerous situation, typically over-pressure
- Schrader valve: used to hold the air inside automobile tires
- Solenoid valve: an electrically controlled hydraulic or pneumatic valve
- Stopcock: restricts or isolates flow through a pipe
- Swirl valve: A specially designed Joule-Thompson pressure reduction/expansion valve imparting a centrifugal force upon the discharge stream for improving gas-liquid phase separation
- Tap (British English), faucet (American English): the common name for a valve used in homes to regulate water flow
- Thermal expansion valve, used in air conditioning and refrigeration systems.
- Thermostatic Mixing Valve
- Thermostatic Radiator Valve
- Trap primer: sometimes include other types of valves, or are valves themselves
- Vacuum breaker valve: prevents the back-siphonage of contaminated water into pressurized drinkable water supplies.

In Biology

Nature has developed an efficient valve, the sphincter, found in many animals including humans.

Body

The valve's body is the outer casing of most or all of the valve that contains the internal parts or *trim*. The bonnet is the part of the encasing through which the stem passes and that forms a guide and seal for the stem. The bonnet typically screws into or is bolted to the valve body.

Valve bodies are usually metallic or plastic. Brass, bronze, gunmetal, cast iron, steel, alloy steels and stainless steels are very common. Seawater applications, like desalination plants, often use duplex valves, as well as super duplex valves, due to their corrosion resistant properties, particularly against warm seawater. Alloy 20 valves are typically used in sulphuric acid plants, whilst monel valves are used in hydrofluoric acid (HF Acid) plants. Hastelloy valves are often used in high temperature applications, such as nuclear plants, whilst inconel valves are often used in hydrogen applications. Plastic bodies are used for relatively low pressures and temperatures. PVC, PP, PVDF and glass-reinforced nylon are common plastics used for valve bodies.

Components

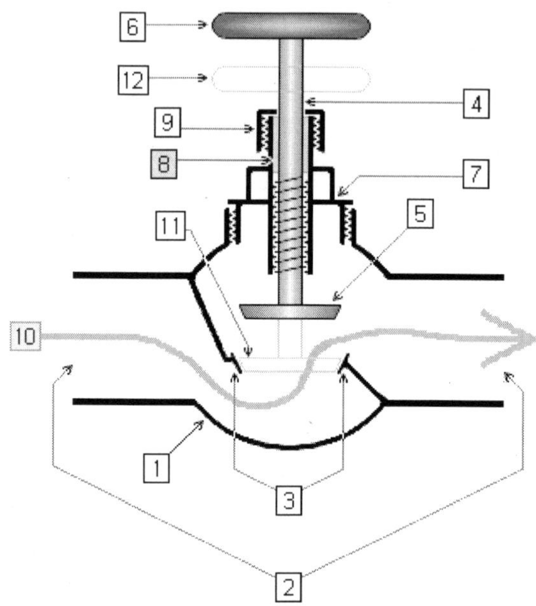

Figure: *Cross-sectional diagram of an open globe valve. 1. body; 2. ports; 3. seat; 4. stem; 5. disc when valve is open; 6. handle or handwheel when valve is open; 7. bonnet; 8. packing; 9. gland nut; 10. fluid flow when valve is open; 11. position of disc if valve were shut; 12. position of handle or handwheel if valve were shut.*

The main parts of the most usual type of valve are the body and the bonnet. These two parts form the casing that holds the fluid going through the valve.

Bonnet

A bonnet acts as a cover on the valve body. It is commonly semi-permanently screwed into the valve body or bolted onto it. During manufacture of the valve, the internal parts are put into the body and then the bonnet is attached to hold everything together inside. To access internal parts of a valve, a user would take off the bonnet, usually for maintenance. Many valves do not have bonnets; for example, plug valves usually do not have bonnets. Many ball valves do not have bonnets since the valve body is put together in a different style, such as being screwed together at the middle of the valve body.

Ports

Ports are passages that allow fluid to pass through the valve. Ports are obstructed by the valve member or disc to control flow. Valves most commonly have 2 ports, but may have as many as 20. The valve is almost always connected at its ports to pipes or other components. Connection methods include threadings, compression fittings, glue, cement, flanges, or welding.

Handle or Actuator

A handle is used to manually control a valve from outside the valve body. Automatically controlled valves often do not have handles, but some may have a handle (or something similar) anyway to manually override automatic control, such as a stop-check valve. An actuator is a mechanism or device to automatically or remotely control a valve from outside the body. Some valves have neither handle nor actuator because they automatically control themselves from inside; for example, check valves and relief valves may have neither.

Disc

A disc or valve member is a movable obstruction inside the stationary body that adjustably restricts flow through the valve. Although traditionally disc-shaped, discs come in various shapes. Depending on the type of valve, a disc can move linearly inside a valve, or rotate on the stem (as in a butterfly valve), or rotate on a hinge or trunnion (as in a check valve). A *ball* is a round valve member with one or more paths between ports passing through it. By rotating the ball, flow can be directed between different ports. Ball valves use spherical rotors with a cylindrical hole drilled as a fluid passage. Plug valves use cylindrical or conically tapered rotors called plugs. Other round shapes for rotors are possible as well in rotor valves, as long as the rotor can be turned inside the valve body. However not all round or spherical discs are rotors; for example, a ball check valve uses the ball to block reverse flow, but is not a rotor because operating the valve does not involve rotation of the ball.

Seat

The seat is the interior surface of the body which contacts the disc to form a leak-tight seal. In discs that move linearly or swing on a hinge or trunnion, the disc comes into contact with the seat only when the valve is shut. In disks that rotate, the seat is always in contact with the disk, but the area of contact changes as the disc is turned. The seat always remains stationary relative to the body.

Seats are classified by whether they are cut directly into the body, or if they are made of a different material:

- Hard seats are integral to the valve body. Nearly all hard seated metal valves have a small amount of leakage.
- Soft seats are fitted to the valve body and made of softer materials such as PTFE or various elastomers such as NBR, EPDM, or FKM depending on the maximum operating temperature.

Stem

The stem transmits motion from the handle or controlling device to the disc. The stem typically passes through the bonnet when present. In some cases, the stem and the disc can be combined in one piece, or the stem and the handle are combined in one piece.

The motion transmitted by the stem may be a linear force, a rotational torque, or some combination of these(Angle valve using torque reactor pin and Hub Assembly). The valve and stem can be threaded such that the stem can be screwed into or out of the valve by turning it in one direction or the other, thus moving the disc back or forth inside the body. Packing is often used between the stem and the bonnet to maintain a seal. Some valves have no external control and do not need a stem as in most check valves.

Valves whose disc is between the seat and the stem and where the stem moves in a direction into the valve to shut it are normally-seated or front seated. Valves whose seat is between the disc and the stem and where the stem moves in a direction out of the valve to shut it are reverse-seated or back seated. These terms don't apply to valves with no stem or valves using rotors.

Gaskets

Gaskets are the mechanical seals, or packings, used to prevent the leakage of a gas or fluids from valves.

Valve Balls

A valve ball is also used for severe duty, high-pressure, high-tolerance applications. They are typically made of stainless steel, titanium, Stellite, Hastelloy, brass, or nickel. They can also be made of different types of plastic, such as ABS, PVC, PP or PVDF.

Spring

Many valves have a spring for spring-loading, to normally shift the disc into some position by default but allow control to reposition the disc. Relief valves commonly use a spring to keep the valve shut, but allow excessive pressure to force the valve open against the spring-loading. Coil springs are normally used. Typical spring materials include zinc plated steel, stainless steel, and for high temperature applications Inconel X750.

Trim

The internal elements of a valve are collectively referred to as a valve's trim. According to API Standards 600, "Steel Gate Valve-Flanged and Butt-welding Ends, Bolted Bonnets", the trim consists of stem, seating surface in the body, gate seating surface, bushing or a deposited weld for the backseat and stem hole guide, and small internal parts that normally contact the service fluid, excluding the pin that is used to make a stem-to-gate connection (this pin shall be made of an austenitic stainless steel material).

Valve Operating Positions

Valve positions are operating conditions determined by the position of the disc or rotor in the valve. Some valves are made to be operated in a gradual change between two or more positions. Return valves and non-return valves allow fluid to move in 2 or 1 directions respectively.

Two-Port Valves

Operating positions for 2-port valves can be either shut (closed) so that no flow at all goes through, fully open for maximum flow, or sometimes partially open to any degree in between. Many valves are not designed to precisely control intermediate degree of flow; such valves are considered to be either open or shut. Some valves are specially designed to regulate varying amounts of flow. Such valves have been called by various names such as *regulating, throttling, metering*, or *needle valves*. For example, needle valves have elongated conically-tapered discs and matching seats for fine flow control. For some valves, there may be a mechanism to indicate by how much the valve is open, but in many cases other indications of flow rate are used, such as separate flow meters.

In plants with remote-controlled process operation, such as oil refineries and petrochemical plants, some 2-way valves can be designated as normally closed (NC) or normally open (NO) during regular operation. Examples of normally closed valves are *sampling valves*, which are only opened while a sample is taken. Examples of normally open valves are *isolation valves*, which are usually only shut when there is a problem with a unit or a section of a fluid system such as a leak in order to isolate the problem from the rest of the system. Although many 2-way valves are made in which the flow can go in either direction between the two ports, when a valve is placed into a certain application, flow is often expected to go from one certain port on the upstream side of the valve, to the other port on the downstream side. Pressure regulators are variations of valves in which flow is controlled to produce a certain downstream pressure, if possible. They are often used to control flow of gas from a gas cylinder. A back-pressure regulator is a variation of a valve in which flow is controlled to maintain a certain upstream pressure, if possible.

Three-Port Valves

Three-way ball valves come with a T-or L-shaped fluid passageways inside the rotor. The T valve might be used to permit connection of one inlet to either or both outlets or connection of the two outlets. The L valve could be used to permit disconnection of both or connection of either but not both of two inlets to one outlet.

Shuttle valves automatically connect the higher pressure inlet to the outlet while (in some configurations) preventing flow from one inlet to the other.

Single handle mixer valves produce a variable mixture of hot and cold water at a variable flow rate under control of a single handle.

Thermostatic mixing valves mix hot and cold water to produce a constant temperature in the presence of variable pressures and temperatures on the two input ports.

Four-Port Valves

The four-way valve or four-way cock is a fluid control valve whose body has four ports equally spaced round the valve chamber and the plug has two passages to connect adjacent ports. The plug may be cylindrical or tapered, or a ball.

It has two flow positions as shown, and usually a central position where all ports are closed.

It can be used to isolate and to simultaneously bypass a sampling cylinder installed on a pressurized water line. It is useful to take a fluid sample without affecting the pressure of a hydraulic system and to avoid degassing (no leak, no gas loss or air entry, no external contamination).

It was used to control the flow of steam to the cylinder of early double-acting steam engines, such as those designed by Richard Trevithick. This use of the valve is possibly attributable to Denis Papin.

Because the two "L"-shaped ports in the plug do not interconnect, the four-way valve is sometimes referred to as an "×" port.

It can be used to isolate and to simultaneously bypass a sampling cylinder installed on a pressurized water line. It is useful to take a fluid sample without affecting the pressure of a hydraulic system and to avoid degassing (no leak, no gas loss or air entry, no external contamination)....

Control

Many valves are controlled manually with a handle attached to the stem. If the handle is turned ninety degrees between operating positions, the valve is called a quarter-turn valve. Butterfly, ball valves, and plug valves are often quarter-turn valves. If the handle is circular with the stem as the axis of rotation in the center of the circle, then the handle is called a handwheel. Valves can also be controlled by actuators attached to the stem. They can be electromechanical actuators such as an electric motor or solenoid, pneumatic actuators which are controlled by air pressure, or hydraulic actuators which are controlled by the pressure of a liquid such as oil or water. Actuators can be used for the purposes of automatic control such as in washing machine cycles, remote control such as the use of a centralised control room, or because manual control is too difficult such as when the valve is very large. Pneumatic actuators and hydraulic actuators need pressurised air or liquid lines to supply the actuator: an inlet line and an outlet line. Pilot valves are valves which are used to control other valves. Pilot valves in the actuator lines control the supply of air or liquid going to the actuators.

The fill valve in a toilet water tank is a liquid level-actuated valve. When a high water level is reached, a mechanism shuts the valve which fills the tank.

In some valve designs, the pressure of the flow fluid itself or pressure difference of the flow fluid between the ports automatically controls flow through the valve.

Other Considerations

Valves are typically rated for maximum temperature and pressure by the manufacturer. The wetted materials in a valve are usually identified also. Some valves rated at very high pressures are available.

When a designer, engineer, or user decides to use a valve for an application, he/she should ensure the rated maximum temperature and pressure are never exceeded and that the wetted materials are compatible with the fluid the valve interior is exposed to. In Europe, valve design and pressure ratings are subject to statutory regulation under the Pressure Equipment Directive 97/23/EC (PED)

Some fluid system designs, especially in chemical or power plants, are schematically represented in piping and instrumentation diagrams. In such diagrams, different types of valves are represented by certain symbols.

Valves in good condition should be leak-free. However, valves may eventually wear out from use and develop a leak, either between the inside and outside of the valve or, when the valve is shut to stop flow, between the disc and the seat. A particle trapped between the seat and disc could also cause such leakage.

Chapter 6

New Technology

A pipe is a tubular section or hollow cylinder, usually but not necessarily of circular cross-section, used mainly to convey substances which can flow — liquids and gases (fluids), slurries, powders, masses of small solids. It can also be used for structural applications; hollow pipe is far stiffer per unit weight than solid members.

Figure: *Steel pipes*

In common usage the words *pipe* and *tube* are usually interchangeable, but in industry and engineering, the terms are uniquely defined. Depending on the applicable standard to which it is manufactured, pipe is generally specified by a nominal diameter with a constant outside diameter (OD) and a schedule that defines the thickness. Tube is most often specified by the OD and wall thickness, but may be specified by any two of OD, inside diameter (ID), and wall thickness. Pipe is generally manufactured to one of several international and national industrial standards. While similar standards exist for specific industry application tubing, tube is often made to custom sizes and a broader range of diameters and tolerances. Many industrial and government standards exist for the production of pipe and tubing. The term "tube" is also commonly applied to non-cylindrical sections, i.e., square or rectangular tubing. In general, "pipe" is the more common term in most of the world, whereas "tube" is more widely used in the United States.

Figure: *Plastic (PVC) pipes in USA*

Both "pipe" and "tube" imply a level of rigidity and permanence, whereas a *hose* (or hosepipe) is usually portable and flexible. Pipe assemblies are almost always constructed with the use of fittings such as elbows, tees, and so on, while tube may be formed or bent into custom configurations. For materials that are inflexible, cannot be formed or where construction is governed by codes or standards, tube assemblies are also constructed with the use of tube fittings.

Uses

- Domestic water systems
- Pipelines transporting gas or liquid over long distances
- Scaffolding.

Scaffolding is a temporary structure used to support people and material in the construction or repair of buildings and other large structures. It is usually a modular system of metal pipes or tubes, although it can be from other materials. Bamboo is frequently used in some Asian countries, such as Hong Kong.

Scaffolding in the Modern Day

This European Standard specifies performance requirements and methods of structural and general design for access and working scaffolds. Requirements given are for scaffold structures that rely on the adjacent structures for stability. In general these requirements also apply to other types of working scaffolds.

The purpose of a working scaffold is to provide a safe place of work with safe access suitable for the work being done. This document sets out performance requirements for working scaffolds. These are substantially independent of the materials of which the scaffold is made. The standard is intended to be used as the basis for enquiry and design.

Thus the requirements of BS EN 12811-1. TG20 is largely based on BS 5973 with extracts taken directly from the old code, it also uses permissible stress design method. However, TG20 received a mixed response from the UK industry and as a result TG20 is being re-written and the new version is due for release sometime in 2008. This is the reason for the 'limbo' situation. Until the release of the revised TG20 the HSE continue to allow scaffold to be built in accordance with BS 5973.

Materials: The basic materials are tubes, couplers and boards.

Tubes are either steel or aluminium, although composite scaffolding uses filament wound tubes of glass fibre in a nylon or polyester matrix. If steel they are either 'black' or galvanised. The tubes come in a variety of lengths and a standard diameter of 48.3 mm. (1.5 NPS pipe). The chief difference between the two types of tubes is the lower weight of aluminium tubes (1.7 kg/m as opposed to 4.4 kg/m) and also a greater flexibility and so less resistance to force. Tubes are generally bought in 6.3 m lengths and can then be cut down to certain typical sizes.

Boards provide a working surface for users of the scaffold. They are seasoned wood and come in three thicknesses (38 mm (usual), 50 mm and 63 mm) are a standard width (225 mm) and are a maximum of 3.9 m long. The board ends are protected by metal plates called hoop irons or sometimes nail plates. Timber Scaffold boards in the UK should comply with the requirements of BS 2482. As well as timber, steel or aluminium decking is used or laminate boards. As well as boards for the working platform there are sole boards which are placed beneath the scaffolding if the surface is soft or otherwise suspect, although ordinary boards can be used, another design called the scaffpad is another solution as it is made from a rubber base with a base plate moulded inside, these are great to put on uneven ground because they adapt to any ground where sole boards would split costing more money to replace.

Couplers are the fittings which hold the tubes together. The most common are called scaffold couplers, there are three basic types: *right-angle couplers*, *putlog couplers* and *swivel couplers*. To join tubes end-to-end *joint pins* (also called spigots) or *sleeve couplers* are used, or both together. Only right angle couplers and swivel couplers can be used to fix tube in a 'load-bearing connection'. Single couplers are not load-bearing couplers and have no design capacity.

Other common materials lnclude base plates, ladders, ropes, anchor ties, reveal ties, gin wheels, sheeting, etc.

Despite the metric measurements given many scaffolders measure tubes and boards in imperial units. With tubes from 21 feet down and boards from 13 ft down.

Bamboo scaffolding is widely used in Hong Kong, with nylon straps tied into knots as couplers.

Basic Scaffolding: The key elements of a scaffold are *standards*, *ledgers* and *transoms*. The standards, also called uprights, are the vertical tubes that transfer the entire mass of the structure to the ground where they rest on a square *base plate* to spread the load. The base plate has a shank in its centre to hold the tube and is sometimes pinned to a *sole board*. Ledgers are horizontal tubes which connect between the standards. Transoms rest upon the ledgers at right angles. *Main transoms* are placed next to the standards, they hold the standards in place and provide support for boards; *intermediate transoms* are those placed between the main transoms to provide extra support for boards. In Canada this style is referred to as "English". "American" has the transoms attached to the standards and is used less but has certain advantages in some situations. Since scaffolding is a physical structure, it is possible to go in and come out of scaffolding.

As well as the tubes at right angles there are *cross braces* to increase rigidity, these are placed diagonally from ledger to ledger, next to the standards to which they are fitted. If the braces are fitted to the ledgers they are called ledger braces. To limit sway a *facade brace* is fitted to the face of the scaffold every 30 metres or so at an angle of 35°-55° running right from the base to the top of the scaffold and fixed at every level.

Of the couplers previously mentioned, right-angle couplers join ledgers or transoms to standards, putlog or single couplers join board bearing transoms to ledgers-Non-board bearing transoms should be fixed using a right-angle coupler. Swivel couplers are to connect tubes at any other angle. The actual joints are staggered to avoid occurring at the same level in neighbouring standards.

The spacings of the basic elements in the scaffold are fairly standard. For a general purpose scaffold the maximum bay length is 2.1 m, for heavier work the bay size is reduced to 2 or even 1.8 m while for inspection a bay width of up to 2.7 m is allowed.

The scaffolding width is determined by the width of the boards, the minimum width allowed is 600 mm but a more typical four-board scaffold would be 870 mm wide from standard to standard. More heavy-duty scaffolding can require 5, 6 or even up to 8 boards width. Often an *inside board* is added to reduce the gap between the inner standard and the structure.

The lift height, the spacing between ledgers, is 2 m, although the base lift can be up to 2.7 m.

Transom spacing is determined by the thickness of the boards supported, 38 mm boards require a transom spacing of no more than 1.2 m while a 50 mm board can stand a transom spacing of 2.6 m and 63 mm boards can have a maximum span of 3.25 m. The minimum overhang for all boards is 50 mm and the maximum overhang is no more than 4x the thickness of the board.

Foundations: Good foundations are essential. Often scaffold frameworks will require more than simple base plates to safely carry and spread the load. Scaffolding can be used without base plates on concrete or similar hard surfaces, although base plates are always recommended. For surfaces like pavements or tarmac base plates are necessary. For softer or more doubtful surfaces sole boards must be used, beneath a single standard a sole board should be at least 1,000 cm^2 with no dimension less than 220 mm, the thickness must be at least 35 mm. For heavier duty scaffold much more substantial baulks set in concrete can be required. On uneven ground steps must be cut for the base plates, a minimum step size of around 450 mm is recommended. A working platform requires certain other elements to be safe. They must be close-boarded, have double guard rails and toe and stop boards. Safe and secure access must also be provided.

Ties: Scaffolds are only rarely independent structures. To provide stability for a scaffolding (at left) framework ties are generally fixed to the adjacent building/fabric/steelwork.

General practice is to attach a tie every 4m on alternate lifts (traditional scaffolding). Prefabricated System scaffolds require structural connections at all frames-ie.2-3m centres (tie patterns must be provided by the System manufacturer/supplier). The ties are coupled to the scaffold as close to the junction of standard and ledger (node point) as possible. Due to recent regulation changes, scaffolding ties must support +/-loads (tie/butt loads) and lateral (shear) loads.

Due to the different nature of structures there is a variety of different ties to take advantage of the opportunities.

Through ties are put through structure openings such as windows. A vertical inside tube crossing the opening is attached to the scaffold by a transom and a crossing horizontal tube on the outside called a bridle tube. The gaps between the tubes and the structure surfaces are packed or wedged with timber sections to ensure a solid fit.

Box ties are used to attach the scaffold to suitable pillars or comparable features. Two additional transoms are put across from the lift on each side of the feature and are joined on both sides with shorter tubes called tie tubes. When a complete box tie is impossible a l-shaped *lip tie* can be used to hook the scaffold to the structure, to limit inward movement an additional transom, a *butt transom*, is place hard against the outside face of the structure. Sometimes it is possible to use *anchor ties* (also called *bolt ties*), these are ties fitted into holes drilled in the structure. A common type is a ring bolt with an expanding wedge which is then tied to a node point.

The least 'invasive' tie is a *reveal tie*. These use an opening in the structure but use a tube wedged horizontally in the opening. The reveal tube is usually held in place by a reveal screw pin (an adjustable threaded bar) and protective packing at either end. A transom tie tube links the reveal tube to the scaffold. Reveal ties are not well regarded, they rely solely on friction and need regular checking so it is not recommended that more than half of all ties be reveal ties.

If it is not possible to use a safe number of ties *rakers* can be used. These are single tubes attached to a ledger extending out from the scaffold at an angle of less than 75° and securely founded. A transom at the base then completes a triangle back to the base of the main scaffold.

Putlog Scaffold: As well as putlog couplers there are also putlog tubes, these have a flattened end or have been fitted with a blade. This feature allows the end of the tube to be within or rest upon the brickwork of the structure. They can be called a bricklayer's scaffold and as such consist only of a single row of standards with a single ledger, the putlogs are transoms-attached to the ledger at one end but integrated into the bricks at the other. Spacing is as general purpose scaffold and ties are still required.

The widespread use of scaffolding systems, along with the profound importance that they earned in modern applications such as civil engineering projects and temporary structures, led to the definition of a series of standards covering a vast number of specific issues involving scaffolding. Among the standards there are:

- DIN 4420, a DIN standard divided in 5 parts which covers the design and detail of scaffolds, ladder scaffolds, safety requirements and standard types, materials, components, dimensions and loadbearing capacity.
- DIN 4421, a DIN standard which covers the analysis, design and construction of falsework
- Structural steel
- As components in mechanical systems such as:
    - o Rollers in conveyor belts
    - o Compactors (E.g.: steam rollers)
    - o Bearing casing
- Casing for concrete pilings used in construction projects
- High temperature or pressure manufacturing processes
- The petroleum industry:

o Oil well casing

o Oil refinery equipment

- Delivery of fluids, either gaseous or liquid, in a process plant from one point to another point in the process
- Delivery of bulk solids, in a food or process plant from one point to another point in the process
- The construction of high pressure storage vessels (note that large pressure vessels are constructed from plate, not pipe owing to their wall thickness and size).

Manufacture

There are three processes for metallic pipe manufacture. Centrifugal casting of hot alloyed metal is one of the most prominent process. Ductile iron pipes are generally manufactured in such a fashion. Seamless (SMLS) pipe is formed by drawing a solid billet over a piercing rod to create the hollow shell. Seamless pipe withstands pressure better than other types, and is often more easily available than welded pipe. Welded pipe is formed by rolling plate and welding the seam. The weld flash can be removed from the outside or inside surfaces using a scarfing blade. The weld zone can also be heat treated to make the seam less visible. Welded pipe often has tighter dimensional tolerances than seamless, and can be cheaper if manufactured in the same quantities. Large-diameter pipe (25 centimetres (10 in) or greater) may be ERW, EFW or Submerged Arc Welded ("SAW") pipe.

Tubing, either metal or plastic, is generally extruded.

Materials

Pipe are made in many materials including ceramic, fiberglass, many metals, concrete and plastic. In the past wood and lead (Latin *plumbum*, from which we get the word plumbing) were commonly used. Metallic pipes are commonly made from steel or iron; the finish and metal chemistry are peculiar to the use, fit and form. Typically metallic piping is made of steel or iron, such as unfinished, black (lacquer) steel, carbon steel, stainless steel or galvanized steel, brass, and ductile iron.

Figure: *Historic water mains from Philadelphia included wooden pipes*

Aluminium pipe or tubing may be utilized where iron is incompatible with the service fluid or where weight is a concern; aluminium is also used for heat transfer tubing such as in refrigerant systems. Copper tubing is popular for domestic water (potable) plumbing systems; copper may be used where heat transfer is desirable (i.e. radiators or heat exchangers). Inconel, chrome moly, and titanium steel alloys are used in high temperature and pressure piping in process and power facilities. When specifying alloys for new processes, the known issues of creep and sensitization effect must be taken into account.

Lead piping is still found in old domestic and other water distribution systems, but it is no longer permitted for new potable water piping installations due to its toxicity. Many building codes now require that lead piping in residential or institutional installations be replaced with non-toxic piping or that the tubes' interiors be treated with phosphoric acid. According to a senior researcher and lead expert with the Canadian Environmental Law Association, "...there is no safe level of lead [for human exposure]".

Plastic tubing is widely used for its light weight, chemical resistance, non-corrosive properties, and ease of making connections. Plastic materials include polyvinyl chloride (PVC), chlorinated polyvinyl chloride (CPVC), fibre reinforced plastic (FRP), reinforced polymer mortar (RPMP), polypropylene (PP), polyethylene (PE), cross-linked high-density polyethylene (PEX), polybutylene (PB), and acrylonitrile butadiene styrene (ABS), for example. In many countries, PVC pipes account for most pipe materials used in buried municipal applications for drinking water distribution and wastewater mains.

Pipe may be made from concrete or ceramic, usually for low-pressure applications such as gravity flow or drainage. Pipes for sewage are still predominantly made from concrete or vitrified clay. In other applications plastic pipes provide better properties and processing options.

Reinforced concrete can be used for large-diameter concrete pipes. This pipe material can be used in many types of construction, and is often used in the gravity-flow transport of storm water. Usually such pipe will have a receiving bell or a stepped fitting, with various sealing methods applied at installation.

Sizes

Pipe sizes can be confusing because the terminology may relate to historical dimensions. For example, a half-inch iron pipe does not have any dimension that is a half inch. Initially, a half inch pipe did have an inner diameter of 0.5 inches (13 mm)—but it also had thick walls. As technology improved, thinner walls became possible, but the outside diameter stayed the same so it could mate with existing older pipe, increasing the inner diameter beyond half an inch. The history of copper pipe is similar. In the 1930s, the pipe was designated by its internal diameter and a ^{1}D$_{16}$-inch (1.6 mm) wall thickness. Consequently, a 1-inch (25 mm) copper pipe had a 1^{1}D$_{8}$-inch (28.58 mm) outside diameter. The outside diameter was the important dimension for mating with fittings. The wall thickness on modern copper is usually thinner than ^{1}D$_{16}$ inches (1.6 mm), so the internal diameter is only "nominal" rather than a controlling dimension. Newer pipe technologies sometimes adopted a sizing system as its own. PVC pipe uses the Nominal Pipe Size.

Nominal Pipe Size (NPS) is a North American set of standard sizes for pipes used for high or low pressures and temperatures. Pipe size is specified with two non-dimensional numbers: a nominal pipe size (NPS) for diameter based on inches, and a schedule (Sched. or Sch.) for wall thickness. NPS is often incorrectly called National Pipe Size, due to confusion with national pipe thread (NPT). The European designation equivalent to NPS is *DN* (diamètre nominal/nominal diameter), in which sizes are measured in millimetres. The term NB (nominal bore) is also frequently used interchangeably with NPS.

History

In March 1927, the American Standards Association authorized a committee to standardize the dimensions of wrought steel and wrought iron pipe and tubing. At that time only a small selection of wall thicknesses were in use: standard weight (STD), extra-strong (XS), and double extra-strong (XXS), based on the iron pipe size (IPS) system of the day. However these three sizes did not fit all applications. The committee surveyed the industry and created a system of schedule numbers that designated wall thicknesses based on smaller steps between sizes, although IPS and NPS numbers remain equivalent. The original intent was that each schedule would relate to a given pressure rating, however the numbers deviated so far from wall thicknesses in common use that this original intent could not be accomplished. Also, in 1939, it was hoped that the designations of STD, XS, and XXS would be phased out by schedule numbers, however those original terms are still in common use today (although sometimes referred to as *standard, extra-heavy (XH),* and *double extra-heavy (XXH)*, respectively). Since the original schedules were created, there have been many revisions and additions to the tables of pipe sizes based on industry use and on standards from API, ASTM, and others.

Stainless steel pipes, which were coming into more common use in the mid 20th century, permitted the use of thinner pipe walls with much less risk of failure due to corrosion. By 1949 thinner schedules

5S and 10S, which were based on the pressure requirements modified to the nearest BWG number, had been created, and other "S" sizes followed later. Due to their thin walls, the smaller "S" sizes can not be threaded together according to ASME code, but must be fusion welded.

Application

Based on the NPS and schedule of a pipe, the pipe outside diameter (OD) and wall thickness can be obtained from reference tables such as those below, which are based on ASME standards B36.10M and B36.19M. For example, NPS 14 Sch 40 has an OD of 14 inches and a wall thickness of 0.437 inches. However the NPS and OD values are not always equal, which can create confusion.

- For NPS 1/s to 12 inches, the NPS and OD values are different. For example, the OD of an NPS 12 pipe is actually 12.75 inches. To find the actual OD for each NPS value.
- For NPS 14 inches and up, the NPS and OD values are equal. In other words, an NPS 14 pipe is actually 14 inches OD.

The reason for the discrepancy for NPS 1/s to 12 inches is that these NPS values were originally set to give the same *inside* diameter (ID) based on wall thicknesses standard at the time. However, as the set of available wall thicknesses evolved, the ID changed and NPS became only indirectly related to ID and OD.

For a given NPS, the OD stays fixed and the wall thickness increases with schedule. For a given schedule, the OD increases with NPS while the wall thickness stays constant or increases. Using equations and rules in ASME B31.3 Process Piping, it can be shown that pressure rating decreases with increasing NPS and constant schedule.

Some specifications use pipe schedules called standard wall (STD), extra strong (XS), and double extra strong (XXS), although these actually belong to an older system called iron pipe size (IPS). The IPS number is the same as the NPS number. STD is identical to SCH 40S, and 40S is identical to 40 for NPS 1/8 to NPS 10, inclusive. XS is identical to SCH 80S, and 80S is identical to 80 for NPS 1/8 to NPS 8, inclusive. XXS wall is thicker than schedule 160 from NPS 1/8" to NPS 6" inclusive, and schedule 160 is thicker than XXS wall for NPS 8" and larger.

The "S" designation, for example "NPS Sch 10S", most often indicates stainless steel pipes. However some stainless steel pipes are available in steel designations, so strictly speaking the "S" designation only differentiates B36.19M pipe from B36.10M pipe.

Both polyvinyl chloride pipe (PVC) and chlorinated polyvinyl chloride pipe (CPVC) are made in NPS sizes.

There are two common methods for designating pipe outside diameter (OD). The North American method is called NPS ("Nominal Pipe Size") and is based on inches (also frequently referred to as NB ("Nominal Bore")). The European version is called DN ("Diametre Nominal"/"Nominal Diameter") and is based on millimetres. Designating the outside diameter allows pipes of the same size to be fit together no matter what the wall thickness.

- For pipe sizes less than NPS 14 inch (DN 350), both methods give a nominal value for the OD that is rounded off and is not the same as the actual OD. For example, NPS 2 inch and DN 50 are the same pipe, but the actual OD is 2.375 inches or 60.33 millimetres. The only way to obtain the actual OD is to look it up in a reference table.
- For pipe sizes of NPS 14 inch (DN 350) and greater the NPS size is the actual diameter in inches and the DN size is equal to NPS times 25 (not 25.4) rounded to a convenient multiple of 50. For example, NPS 14 has an OD of 14 inches or 355.60 millimetres, and is equivalent to DN 350.

Since the outside diameter is fixed for a given pipe size, the inside diameter will vary depending on the wall thickness of the pipe. For example, 2" Schedule 80 pipe has thicker walls and therefore a smaller inside diameter than 2" Schedule 40 pipe. Steel pipe has been produced for about 150 years. The pipe sizes that are in use today in PVC and galvanized were originally designed years ago for steel pipe. The number system, like Sch 40, 80, 160, were set long ago and seem a little odd. For example, Sch 20 pipe is even thinner than Sch 40, but same OD. And while these pipes are based on old steel pipe sizes, there is other pipe, like gold-flow cpvc for heated water, that uses pipe sizes, inside and out, based on old copper pipe size standards instead of steel.

Many different standards exist for pipe sizes, and their prevalence varies depending on industry and geographical area. The pipe size designation generally includes two numbers; one that indicates the outside (OD) or nominal diameter, and the other that indicates the wall thickness. In the early twentieth century, American pipe was sized by inside diameter. This practice was abandoned to improve compatibility with pipe fittings that must usually fit the OD of the pipe, but it has had a lasting impact on modern standards around the world.

In North America and the UK, pressure piping is usually specified by Nominal Pipe Size (NPS) and schedule (SCH). Pipe sizes are documented by a number of standards, including API 5L, ANSI/ASME B36.10M in the US, and BS 1600 and BS 1387 in the United Kingdom. Typically the pipe wall thickness is the controlled variable, and the Inside Diameter (I.D.) is allowed to vary. The pipe wall thickness has a variance of approximately 12.5 percent.

In the rest of Europe pressure piping uses the same pipe IDs and wall thicknesses as Nominal Pipe Size, but labels them with a metric Diameter Nominal (DN) instead of the imperial NPS. For NPS larger than 14, the DN is equal to the NPS multiplied by 25. (Not 25.4) This is documented by EN 10255 (formerly DIN 2448 and BS 1387) and ISO 65, and it is often called DIN or ISO pipe.

Japan has its own set of standard pipe sizes, often called JIS pipe.

The Iron pipe size (IPS) is an older system still used by some manufacturers and legacy drawings and equipment. The IPS number is the same as the NPS number, but the schedules were limited to Standard Wall (STD), Extra Strong (XS), and Double Extra Strong (XXS). STD is identical to SCH 40 for NPS 1/8 to NPS 10, inclusive, and indicates .375" wall thickness for NPS 12 and larger. XS is identical to SCH 80 for NPS 1/8 to NPS 8, inclusive, and indicates .500" wall thickness for NPS 8 and larger. Different definitions exist for XXS, however it is never the same as SCH 160. XXS is in fact thicker than SCH 160 for NPS 1/8" to 6" inclusive, whereas SCH 160 is thicker than XXS for NPS 8" and larger.

Another old system is the Ductile Iron Pipe Size (DIPS), which generally has larger ODs than IPS.

Copper plumbing tube for residential plumbing follows an entirely different size system in America, often called Copper Tube Size (CTS); see domestic water system. Its nominal size is neither the inside nor outside diameter. Plastic tubing, such as PVC and CPVC, for plumbing applications also has different sizing standards.

Agricultural applications use PIP sizes, which stands for Plastic Irrigation Pipe. PIP comes in pressure ratings of 22 psi (150 kPa), 50 psi (340 kPa), 80 psi (550 kPa), 100 psi (690 kPa), and 125 psi (860 kPa) and is generally available in diameters of 6", 8", 10", 12", 15", 18", 21", and 24".

Standards

The manufacture and installation of pressure piping is tightly regulated by the ASME "B31" code series such as B31.1 or B31.3 which have their basis in the ASME Boiler and Pressure Vessel Code. This code has the force of law in Canada and the USA. Europe and the rest of the world has an equivalent system of codes. Pressure piping is generally pipe that must carry pressures greater than 10 to 25 atmospheres, although definitions vary. To ensure safe operation of the system, the manufacture, storage, welding, testing, etc. of pressure piping must meet stringent quality standards.

Manufacturing standards for pipes commonly require a test of chemical composition and a series of mechanical strength tests for each heat of pipe. A heat of pipe is all forged from the same cast ingot, and therefore had the same chemical composition. Mechanical tests may be associated to a lot of pipe, which would be all from the same heat and have been through the same heat treatment processes. The manufacturer performs these tests and reports the composition in a mill traceability report and the mechanical tests in a material test report, both of which are referred to by the acronym MTR. Material with these associated test reports is called traceable. For critical applications, third party verification of these tests may be required; in this case an independent lab will produce a certified material test report (CMTR), and the material will be called certified.

Some widely used pipe standards are:

- The API range-now ISO 3183. E.g.: API 5L Grade B-now ISO L245 where the number indicates yield strength in MPa
- ASME SA106 Grade B (Seamless carbon steel pipe for high temperature service)
- ASTM A312 (Seamless and welded austenitic stainless steel pipe)
- ASTM C76 (Concrete Pipe)
- ASTM D3033/3034 (PVC Pipe)
- ASTM D2239 (Polyethylene Pipe).

API 5L was changed in the second half of 2008 to edition 44 from edition 43 to make it identical to ISO 3183. It is important to note that the change has created the requirement that sour service, ERW pipe, pass a hydrogen induced cracking (HIC) test per NACE TM0284 in order to be used for sour service.

- ACPA [American Concrete Pipe Association]
- AWWA [American Water Works Association]
- AWWA M45.

Traceability and Positive Material Identification (PMI)

Maintaining the traceability between the material and this paperwork is an important quality assurance issue. QA often requires the heat number to be written on the pipe. Precautions must also be taken to prevent the introduction of counterfeit materials. As a backup to etching/labeling of the material identification on the pipe, Positive Material Identification (PMI) is performed using a handheld device; the device scans the pipe material using an emitted electromagnetic wave (x-ray fluorescence/XRF) and receives a reply that is spectrographically analysed.

Installation

Pipe installation is often more expensive than the material and a variety of specialized tools, techniques, and parts have been developed to assist this. Pipe is usually delivered to a customer or jobsite as either "sticks" or lengths of pipe (typically 20 feet) or they are prefabricated with elbows, tees and valves into a prefabricated pipe spool [A pipe spool is a piece of pre-assembled pipe and fittings, usually prepared in a shop so that installation on the construction site can be more efficient.]. They are usually tagged with a bar code and the ends are capped for protection. The Pipe and Pipe Spools are delivered to a warehouse on a large commercial job and they may be held indoors or in a gridded laydown yard. The Pipe or pipe spool is retrieved, staged, rigged, and then lifted into place. On large process jobs the lift is made using cranes and hoist and other material lifts. They are typically temporarily supported in the steel structure using beam clamps, straps, and small hoists until the Pipe Supports are attached or otherwise secured.

An example of a tool used for installation for a small plumbing pipe (threaded ends) is the pipe wrench. Small pipe is typically not heavy and can be lifted into place by the installation craft labourer.

Joining

Pipes are commonly joined by welding, using threaded pipe and fittings; sealing the connection with a pipe thread compound, Polytetrafluoroethylene (PTFE) Thread seal tape, oakum, or PTFE string, or by using a mechanical coupling. Process piping is usually joined by welding using a TIG or MIG process. The most common process pipe joint is the butt weld. The ends of pipe to be welded must have a certain weld preparation called an End Weld Prep (EWP) which is typically at an angle of 37.5 degrees to accommodate the filler weld metal. The most common pipe thread in North America is the National Pipe Thread (NPT) or the Dryseal (NPTF) version. Other pipe threads include the British standard pipe thread (BSPT), the garden hose thread (GHT), and the fire hose coupling (NST).

Copper pipes are typically joined by soldering, brazing, compression fittings, flaring, or crimping. Plastic pipes may be joined by solvent welding, heat fusion, or elastomeric sealing.

If frequent disconnection will be required, gasketed pipe flanges or union fittings provide better reliability than threads. Some thin-walled pipes of ductile material, such as the smaller copper or flexible plastic water pipes found in homes for ice makers and humidifiers, for example, may be joined with compression fittings.

Underground pipe typically uses a "push-on" gasket style of pipe that compresses a gasket into a space formed between the two adjoining pieces. Push-on joints are available on most types of pipe. A pipe joint lubricant must be used in the assembly of the pipe. Under buried conditions, gasket-joint pipes allow for lateral movement due to soil shifting as well as expansion/contraction due to temperature differentials. Plastic MDPE and HDPE gas and water pipes are also often joined with Electrofusion fittings. Large above ground pipe typically uses a flanged joint, which is generally available in ductile iron pipe and some others. It is a gasket style where the flanges of the adjoining pipes are bolted together, compressing the gasket into a space between the pipe. Mechanical grooved couplings or Victaulic joints are also frequently used for frequent disassembly & assembly. Developed in the 1920s, these mechanical grooved couplings can operate up to 1,200psi working pressures and available in materials to match the pipe grade. Another type of mechanical coupling is a Swagelok brand fitting; this type of compression fitting is typically used on small tubing under 3/4 inch in diameter.

Fittings and Valves

Fittings are also used to split or join a number of pipes together, and for other purposes. A broad variety of standardized pipe fittings are available; they are generally broken down into either a tee, an elbow, a branch, a reducer/enlarger, or a wye. Valves control fluid flow and regulate pressure.

Pipe Support

Pipes are either supported from below or hung from above. These devices are called pipe supports. Supports may be as simple as a pipe "shoe" which is akin to a half of an I-beam welded to the bottom of the pipe; they may be "hung" using a clevis or trapeze type of devices called pipe hangers. Pipe Supports of any kind may incorporate springs, snubbers and/or dampers to compensate for thermal expansion, or to provide vibration isolation, shock control, or vibration excitation of the pipe due to earthquake motion. Some dampers are simply fluid dashpots whereas other dampers may be active hydraulic devices that have sophisticated systems that act to dampen peak displacements due to forcing functions. The forcing functions may be process derived (such as in a fluidized bed reactor) or from a natural phenomenon such as an earthquake (design basis event or DBE).

Cleaning

The inside of pipes can be cleaned with the tube cleaning process, if they are contaminated with debris or fouling. This depends on the process that the pipe will be used for and the cleanliness needed for the process. In some cases the pipes are cleaned using a displacement device formally known as a Pipeline Inspection Gauge or "pig"; alternately the pipes or tubes may be chemically flushed using specialized solutions that are pumped through. In some cases, where care has been taken in the manufacture, storage, and installation of pipe and tubing, the lines are blown clean with compressed air or nitrogen.

Other Uses

Pipe is widely used in the fabrication of handrails, guardrails and railings.

AN Thread

The AN thread is a particular type of fitting used to connect flexible hoses and rigid metal tubing that carry fluid. It is a US military-derived specification that dates back to World War II and stems from a joint standard agreed upon by the Army and Navy, hence AN. AN sizes range from-2 (dash two) to-32 in irregular steps, with each step equating to the OD (outside diameter) of the tubing in 1/16" increments. Therefore, a-8 AN size would be equal to 1/2" OD tube (8 x 1/16 = 1/2). However, this system does not specify the ID (inside diameter) of the tubing because the tube wall can vary in thickness. Each AN size also uses its own standard thread size.

AN fittings are a flare fitting, using 37° flared tubing to form a metal-metal seal. They are similar to other 37° flared fittings, such as JIC, which is their industrial variant. The two are interchangeable in theory, though this is typically not recommended due to the exacting specifications and demands of the aerospace industry. The differences between them relate to thread class and shape (how tight a fit the threads are), and the metals used.

Note that 37° AN and 45° SAE fittings and tooling are not interchangeable due to the different flaring angles. Mixing them can cause leakage at the flare.

Figure: *A 37° flare type end fitting for flexible hose*

British Standard Brass Thread

British standard brass thread is an imperial unit based screw thread. It adopts the Whitworth thread form with a pitch of 26 threads per inch for all diameters. It is often wrongly called *British Standard Brass* but is not actually covered by a British Standard.

Nominal sizes are usually in the range $^{1}D_{8}$ to 2 inches.

Types of British Standard Pipe Thread

Two types of threads are distinguished:

- Parallel ('straight') threads, *British Standard Pipe Parallel thread* (*BSPP*; originally also known as *British Standard Pipe Fitting thread/BSPF* and *British Standard Pipe Mechanical thread/BSPM*), which have a constant diameter; denoted by the letter *G*.
- Taper threads, *British Standard Pipe Taper thread* (*BSPT*), whose diameter increases or decreases along the length of the thread; denoted by the letter *R*.

These can be combined into two types of joints:

- Jointing threads: These are pipe threads where pressure-tightness is made through the mating of two threads together. They always use a taper male thread, but can have either parallel or taper female threads. (In Europe, taper female pipe threads are not commonly used.)

- Longscrew threads: These are parallel pipe threads used where a pressure-tight joint is achieved by the compression of a soft material (such as an o-ring seal or a washer) between the end face of the male thread and a socket or nipple face, with the tightening of a backnut.

Thread Form

- the threadform follows the British Standard Whitworth standard
- symmetrical V-thread in which the angle between the flanks is 55° (measured in an axial plane)
- one-sixth of this sharp V is truncated at the top and the bottom
- the threads are rounded equally at crests and roots by circular arcs ending tangentially with the flanks
- the theoretical depth of the thread is therefore 0.64 times the nominal pitch

Pipe Thread Sizes

A list of 16 thread sizes are defined by the standards, ranging from $^{1}D_{16}$ to 6. The size number was originally based on the *inner* diameter measured in inches of a steel tube for which the thread was intended, but in the modern metric version of the standard is simply a size number. The major diameter listed is the outer diameter of the external thread. For a taper thread, it is the diameter at the "gauge length" from the small end of the thread. The taper is 1 to 16, meaning that for each 16 units of measurement increase in the distance from the end, the diameter increases by 1 unit of measurement.

Pipe Thread Designations

These standard pipe threads are formally referred to by the following sequence of blocks:

- the words "Pipe thread",
- the document number of the standard (e.g., "ISO 7" or "EN 10226")
- the symbol for the pipe thread type:
    - G = external + internal parallel (ISO 228)
    - R = external taper (ISO 7)
    - Rp = internal parallel (ISO 7/1)
    - Rc = internal taper (ISO 7)
    - Rs = external parallel

- the thread size.

Threads are normally right-hand. For left-hand threads, the letters "LH" are appended.

Example: Pipe thread EN 10226 Rp 2½

The terminology for the use of G and R originated from Germany (G for Gas, as it was originally designed for use on gas pipes ; R for Rohr (meaning pipe).)

Garden Hose Thread

A garden hose is a flexible tube used to carry water. There are a number of common attachments available for the end of the hose, such as sprayers and sprinklers (which are used to concentrate water at one point or spread it over a large area). Hoses are usually attached to a hose spigot (tap).

Figure: *A garden hose*

Garden hoses are commonly green and often black, but can also be found in a variety of other colours. Garden hoses are typically made of synthetic rubber or soft plastic, reinforced with an internal web of fibers. As a result of these materials, garden hoses are flexible and their smooth exterior facilitates pulling them past trees, posts and other obstacles. Garden hoses are also generally tough enough to survive scraping on rocks and being stepped on without damage or leaking. Most garden hoses are not rated for use with hot water, and their packaging will generally specify whether or not this is the case.

Typical hoses used for filling of the potable water tanks in recreational vehicles are similar to garden hoses but are made of nonporous materials that are less likely to collect bacteria or affect the water's taste. They are coloured white as a distinguishing characteristic.

The term hosepipe is chiefly British, South African and southern U.S. usage; garden hose (more commonly just hose) is the predominant term in other English-speaking countries. The term "hose" is also used for other types of flexible, water-carrying tubes such as those used by fire departments.

Standards and Connectors

Garden hoses connect using a male/female thread connection Spigots typically have male connectors and the hose has a captive nut which fits the threads. (The technical term for this arrangement is a hose union).

The thread standard for garden hoses in the US and its territories is known as GHT or "garden hose thread" which has an outer diameter of 1.0625 inches (27.0 mm) and a pitch of 11.5 TPI). Outside the US, the more common BSP standard is used, which is 3/4" and 14 TPI. The GHT and BSP standards are not compatible.

Each end of a typical garden hose mate with one another, which allows multiple garden hoses to be linked end-to-end to increase their length. Small rubber or plastic seals are used in female ends to prevent leakage. Sometimes these seals disintegrate or fall out of older hoses, which results in high-pressure leakage spraying from the hose.

National Pipe Thread

National Pipe Thread Tapered Thread (NPT) is a U.S. standard for tapered threads used on threaded pipes and fittings. The taper rate for all NPT threads is $^{1}D_{16}$ ($^{3}D_{4}$ inch per foot) measured by the change of diameter (of the pipe thread) over distance. The angle between the taper and the center axis of the pipe is 1° 472 243 (1.7899°). Commonly-used sizes are $^{1}D_{8}$, $^{1}D_{4}$, $^{3}D_{8}$, $^{1}D_{2}$, $^{3}D_{4}$, 1, 1 $^{1}D_{4}$, 1 $^{1}D_{2}$, and 2 inch, appearing on pipes and fittings by most U.S. suppliers. Sizes smaller than $^{1}D_{8}$ are occasionally used for compressed air, while sizes larger than 2 in are uncommon, due to the use of alternative methods of joining that are used with these larger sizes. NPT is defined by ANSI/ ASME standard B1.20.1.

Nominal Pipe Size is loosely related to the inside diameter of Schedule 40 pipe. Because of the pipe wall thickness, the actual diameter of the threads is larger than the NPS, considerably so for small NPS. Other schedules of pipe have different wall thickness but the OD (outer diameter) and thread profile remain the same, so the inside diameter of the pipe is therefore different from the nominal diameter.

Threaded pipes can provide an effective seal for pipes transporting liquids, gases, steam, and hydraulic fluid. These threads are now used in materials other than steel and brass, including PTFE, PVC, nylon, bronze and cast iron.

Table: *Pipe Thread Sizes*

Nominal pipe size (in)	*Pipe outer diameter*	*Threads per inch*	*Thread pitch*
$^{1}D_{16}$	0.3125 in (7.94 mm)	27	0.03704 in (0.94082 mm)
$^{1}D_{8}$	0.405 in (10.29 mm)	27	0.03704 in (0.94082 mm)
$^{1}D_{4}$	0.540 in (13.72 mm)	18	0.05556 in (1.41122 mm)
$^{3}D_{8}$	0.675 in (17.15 mm)	18	0.05556 in (1.41122 mm)
$^{1}D_{2}$	0.840 in (21.34 mm)	14	0.07143 in (1.81432 mm)
$^{3}D_{4}$	1.050 in (26.67 mm)	14	0.07143 in (1.81432 mm)
1	1.315 in (33.40 mm)	11 $^{1}D_{2}$	0.08696 in (2.20878 mm)
1 $^{1}D_{4}$	1.660 in (42.16 mm)	11 $^{1}D_{2}$	0.08696 in (2.20878 mm)
1 $^{1}D_{2}$	1.900 in (48.26 mm)	11 $^{1}D_{2}$	0.08696 in (2.20878 mm)
2	2.375 in (60.33 mm)	11 $^{1}D_{2}$	0.08696 in (2.20878 mm)
2 $^{1}D_{2}$	2.875 in (73.03 mm)	8	0.12500 in (3.175 mm)
3	3.500 in (88.90 mm)	8	0.12500 in (3.175 mm)
4	4.500 in (114.30 mm)	8	0.12500 in (3.175 mm)
5	5.563 in (141.30 mm)	8	0.12500 in (3.175 mm)
6	6.625 in (168.28 mm)	8	0.12500 in (3.175 mm)
10	10.750 in (273.05 mm)	8	0.12500 in (3.175 mm)
12	12.750 in (323.85 mm)	8	0.12500 in (3.175 mm)
14	14 in (355.60 mm)	8	0.12500 in (3.175 mm)
16	16 in (406.40 mm)	8	0.12500 in (3.175 mm)
18	18 in (457.20 mm)	8	0.12500 in (3.175 mm)
20	20 in (508.00 mm)	8	0.12500 in (3.175 mm)
24	24 in (609.60 mm)	8	0.12500 in (3.175 mm)

The taper on NPT threads allows them to form a seal when torqued as the flanks of the threads compress against each other, as opposed to parallel/straight thread fittings or compression fittings in which the threads merely hold the pieces together and do not provide the seal. However, a clearance remains between the crests and roots of the threads, resulting in a leakage around this spiral. This means that NPT fittings must be made leak free with the aid of thread seal tape or a thread sealant compound. (The use of tape or sealant will also help to limit corrosion on the threads, which otherwise can make future disassembly nearly impossible.)

There is also a semi-compatible variant called National Pipe Taper Fuel (NPTF) also called Dryseal American National Standard Taper Pipe Thread, defined by ANSI B1.20.3, designed to provide a more leak-free seal without the use of teflon tape or other sealant compound. NPTF threads are the same basic shape but with crest and root heights adjusted for an interference fit, eliminating the spiral leakage path.

Sometimes NPT threads are referred to as MPT ('Male Pipe Thread'), MNPT, or NPT(M) for male (external) threads; and FPT ('Female Pipe Thread'), FNPT, or NPT(F) for female (internal) threads. An equivalent designation is MIP (Male iron pipe) and FIP (Female iron pipe). Also the terms NPS and NPSM are sometimes used to designate a straight, not tapered, thread. (this should not be confused with NPS meaning Nominal Pipe Size)

Thread Form

NPT and NPS threads have a 60° included angle and have a Sellers thread form (flattened peaks and valleys).

Terminology

Sometimes these terms may be used:

- MIP: stands for Male Iron Pipe, or Male International Pipe, or MPT Male Pipe Thread. It is a term for pipe fittings.
- FIP: stands for Female Iron Pipe, or Female International Pipe, or FPT. It is a term for pipe fittings that MIP fittings fit into.

The difference between FIP, MIP and Compression fittings is in the tapering of the thread. FIP has taper of 1:16 (6.25% slope), compression has taper of 13:96(13.5% slope).

A female iron connection has a tapered thread, which thins out to the end of the pipe. As the fitting is tightened, the ever-decreasing thread depth means that the connection becomes watertight. To properly seal the fitting, paste thread sealant or PTFE (polytetrafluoroethylene) tape wrapped around the thread is required.

MIP fittings are usually found in thicker copper pipe, iron pipe, steel pipe, gas pipe, gas stove, gas oven, gas heater, and gas water heater connections.

Panzergewinde

The *Stahlpanzerrohrgewinde* standard for screw threads, more often called by the shorter name *Panzergewinde* (pronunciation), is a technical standard created in Germany and subsequently used in various countries, such as the German-speaking zone (Germany, Switzerland, Austria) and neighboring European countries.

The German name "*Stahlpanzerrohrgewinde*" translates to English as 'steel conduit thread'. The thread is used to join pieces of electrical conduit and cable glands. Alternate stylings of the German name are *Stahl-Panzer-Rohr-Gewinde* and an abbreviated form, *StaPa-Rohr-Gewinde.*

The standard, codified by the *Deutsches Institut für Normung* (DIN, German Institute for Standardization), is DIN 40430. *Panzergewinde* sizes are named with the prefix *PG* plus a nominal number which approximately corresponds to the maximum cable diameter (in millimetres) that can be passed through the conduit.

Because the walls of the conduit are usually relatively thin, the thread depth should not be very large. Thus a thread angle of 80° is used. The *Verband der Elektrotechnik, Elektronik und Informationstechnik* (VDE) (which began as a trade association for standardization in electrical engineering) originally standardized (and named) the thread for use with conduit and cable glands that were made of steel, although today the thread is used with both steel (typically plated with combinations of nickel, zinc, or tin to resist rusting) and polyvinyl chloride (PVC).

Beginning in 2000, the VDE standard for cable glands (VDE 0619) was formally replaced by EN 50262. After a transitional period of several years during which it could still be used, it was replaced in 2003 by a final metric fine thread with 1.5 mm pitch. Similarly, conduit threads were replaced by EN 60423.

Even today, *Panzergewinde* cable glands are still often found on chemical reactors and bioreactors (for example, PG13.5 thread for screwing in sensors) and various other equipment, enclosures, junction boxes, and connectors.

Pronunciation

- *Stahlpanzerrohrgewinde*: SHTAHL-pahntser-rôr-guh-v-n-duh
- *Panzergewinde*: PAHNTSER-guh-v-n-duh
- PG: peh-geh.

Threaded Pipe

A threaded pipe is a pipe with screw-threaded ends for assembly.

Tapered Threads

The threaded pipes used in some plumbing installations for the delivery of gases or fluids under pressure have a threaded section that is slightly conical (in contrast to the cylindrical section commonly found on bolts and leadscrews). This is called a "tapered thread". The seal provided by a threaded pipe joint depends upon the labyrinth seal created by the threads; upon a positive seal between the threads created by the deformation of the threads when they are tightened to the proper torque; and sometimes on the presence of a sealing coating, such as thread seal tape ("Teflon tape"), or a liquid or paste pipe sealant such as pipe dope. Tapered thread joints typically do not include a gasket. Especially precise threads are known as "dry fit" or "dry seal" meaning that no sealant is required for a gas-tight seal. Such threads are needed where the sealant would contaminate or react with the media inside the piping, e.g., oxygen service.

Tapered threaded fittings are sometimes used on plastic piping. Due to the wedging effect of the tapered thread, extreme care must be used to avoid over-tightening the joint. The over-stressed female fitting may split days, weeks, or even years after initial installation. Therefore many municipal plumbing codes restrict the use of threaded plastic pipe fittings.

Straight Threads

Pipes may also be threaded with cylindrical threaded sections, in which case the threads do not themselves provide any sealing function other than some labyrinth seal effect, which may not be enough to satisfy either functional or code requirements. With straight pipe threads, the seal may be provided by an O-ring seated between the shoulder of the male pipe section and an interior surface on the female part.

Cast iron Pipe

Cast iron pipe is a historical pipe which found widespread use as a pressure pipe for transmission of water, gas and sewage, and as a water drainage pipe, during the 19th and 20th centuries. It comprises predominantly a gray cast iron tube and was frequently used uncoated, although later developments did result in various coatings and linings to reduce corrosion and improve hydraulics. Cast iron pipe was gradually superseded by ductile iron pipe, which is a direct development, with most existing manufacturing plants transitioning to the new material during the 1970s and 1980s. There is currently almost no new manufacture of cast iron pipe.

History—Early Use

The first use of cast iron pipe is not recorded. Cast iron tubes were first manufactured in the 14th century in Europe for cannon and it is presumed that similar tubes would have found use as water pipes at the same time, however, it was not until 1455 in Siegerland that the first officially recorded cast iron water pipe was produced for use in Dillenburg Castle (since destroyed).

The oldest extant water pipes date from the 17th century and were installed to distribute water throughout the gardens of the Chateau de Versailles. These amount to some 35 km of pipe, typically 1m lengths with flanged joints. The extreme age of these pipes make them of considerable historical value. Following extensive refurbishment in 2008 by Saint-Gobain PAM, 80% remain original.

Manufacture—Vertically Cast

The first cast iron pipe was produced in horizontal molds which resulted in an uneven distribution of metal around the pipe circumference. In 1845, the first pipe was cast vertically and by the turn of the century, all pipe was manufactured to this method.

Centrifugally Cast

Subsequent to its invention by Dimitri Sensaud deLavaud, a French-Brazilian, in 1918, much cast iron pipe manufacture shifted to the dramatically different technique of centrifugal casting. Modern ductile iron pipe production continues to use this method of casting.

Standardization

The first standardization of cast iron water pipes in Britain occurred in 1917 with the publishing of BS 78. This standard specified a dimensionless nominal size, which approximately corresponded with the internal diameter in inches of the pipe, and four pressure classes, Class A, Class B, Class C and Class D, each with a specified wall thickness and outer diameter. It is noted that the outer diameter is identical between classes with the exception of sizes 12 to 27, where Classes A and B share one diameter and Classes C and D have another, larger diameter. BS 78 was finally superseded when the U.K. harmonised with incompatible European standards, however, the specified outer dimensions continue to remain in effect (albeit in metric form) as the standard pipe outer diameter for ductile iron pipe in Australia and New Zealand through the descendant specification, AS/ NZS 2280.

Copper Pipe

Table: *Copper Tubing Sizes (CTS) for Plumbing*

Nominal size	*Outside diameter (OD) [in (mm)]*	*Inside diameter (ID)[in (mm)]*		
		Type K	*Type L*	*Type M*
$^3/_8$	$^1/_2$ (12.7)	0.402 (10.211)	0.430 (10.922)	0.450 (11.430)
$^1/_2$	$^5/_8$ (15.875)	0.528 (13.411)	0.545 (13.843)	0.569 (14.453)
$^5/_8$	$^3/_4$ (19.05)	0.652 (16.561)	0.668 (16.967)	0.690 (17.526)
$^3/_4$	$^7/_8$ (22.225)	0.745 (18.923)	0.785 (19.939)	0.811 (20.599)
1	$1^1/_8$ (28.575)	0.995 (25.273)	1.025 (26.035)	1.055 (26.797)
$1^1/_4$	$1^3/_8$ (34.925)	1.245 (31.623)	1.265 (32.131)	1.291 (32.791)
$1^1/_2$	$1^5/_8$ (41.275)	1.481 (37.617)	1.505 (38.227)	1.527 (38.786)
2	$2^1/_8$ (53.975)	1.959 (49.759)	1.985 (50.419)	2.009 (51.029)
$2^1/_2$	$2^5/_8$ (66.675)	2.435 (61.849)	2.465 (62.611)	2.495 (63.373)
3	$3^1/_8$ (79.375)	2.907 (73.838)	2.945 (74.803)	2.981 (75.717)

Sizes

Common wall-thicknesses of copper tubing in the USA are "Type K", "Type L" and "Type M":

- Type K has the thickest wall section of the three types of pressure rated tubing and is commonly used for deep underground burial such as under sidewalks and streets, with a suitable corrosion protection coating or continuous polyethylene sleeve as required by code.
- Type L has a thinner pipe wall section, and is used in residential and commercial water supply and pressure applications.
- Type M has the thinnest wall section, and is generally suitable for condensate and other drains, but sometimes illegal for pressure applications, depending on local codes.

Types K and L are generally available in both hard drawn "sticks" and in rolls of soft annealed tubing, whereas type M is usually only available in hard drawn "sticks". In the plumbing trade the size of copper tubing is measured by its nominal diameter (average inside diameter). Some American trades, heating and cooling technicians for instance, use the outside diameter (OD) to designate copper tube sizes. The HVAC tradesman also use this different measurement to try and not confuse water pipe with copper pipe used for the HVAC trade, as pipe used in the air-conditioning trade uses copper pipe that is made at the factory without processing oils that would be incompatible with the oils used to lubricate the compressors in the AC system. The OD of copper tube is ^{1}D $_{8}$th inch larger than its nominal size. Therefore, 1 inch nominal copper tube and 1 ^{1}D $_{8}$th inch ACR tube are exactly the same tube with different size designations. The wall thickness of the tube, never affects the sizing of the tube. Type K ^{1}D $_{2}$ inch nominal tube, is the same size as Type L ^{1}D $_{2}$ inch nominal tube (^{5}D $_{8}$ inch ACR).

Common wall-thicknesses in Europe are "Type X", "Type Y" and "Type Z", defined by the EN 1057 standard.

- Type X is the most common, and is used in above groundservices including drinking water supply, hot and cold water systems, sanitation, central heating and other general purpose applications.
- Type Y is a thicker walled pipe, used for underground works and heavy duty requirements including hot and cold water supply, gas reticulation, sanitary plumbing, heating and general engineering.
- Type Z is a thinner walled pipe, also used for above groundservices including drinking water supply, hot and cold water systems, sanitation, central heating and other general purpose applications.

In the plumbing trade the size of copper tubing is measured by its outside diameter in millimetres. Common sizes are 15 mm and 22 mm. Thin-walled types used to be relatively inexpensive, but since 2002 copper prices have risen considerably due to rising global demand and a stagnant supply.

Double-Walled Pipe

A double-walled pipe is a secondary contained piping system. It is a pipe within a pipe, or encased in an outer covering, with an annulus (interstitial space) between the two diameters. The inner pipe is the primary or carrier pipe and the outer pipe is called the secondary or containment pipe. The great majority of double-walled piping applications involve wastewater, groundwater, and process safety.

U.S. Environmental Protection Agency (EPA) standards, published in 40 CFR Parts 280 and 281, mandate double-walled piping for many below-ground transport systems in wastewater treatment plants and at sanitary or hazardous-waste landfills or remediation sites. These and other regulations affect below-grade transport of hazardous materials in fuel storage systems, tank farms, drainage or runoff from process plants, and some food-processing and related applications. Double-walled containment systems are also used for certain classes of toxic or corrosive chemicals, mostly gases, used in semiconductor fabrication. Double-walled pipes may also be used where a waste water pipe traverses a drinking water catchment area.

Double-walled pipe system may use plastic pressure pipe systems materials of high-density polyethylene (HDPE), polyvinyl chloride (PVC), chlorinated polyvinyl chloride (CPVC), polypropylene (PP), polyvinylidene fluoride (PVDF), and ethylene chlorotrifluoroethylene copolymer (ECTFE). Dual-wall stainless steel tubing is most commonly used in semiconductor plants for the containment of highly toxic gases. A leak detection system in the containment pipe indicates if the carrier pipe is leaking. Double wall pipes are usually hydrostatically tested, or tested to hold and maintain a target gas pressure before the first use.

A different special application of a double-walled pipe is a jacketed pipe, which is used to make high viscosity liquids flow at elevated temperature through the carrier pipe. The outer pipe circulates hot fluids that heat up the interior carrier pipe and its contents in turn.

Ductile Iron Pipe

Ductile iron pipe is a pipe commonly used for potable water distribution. The predominant wall material is ductile iron, a spheroidized graphite cast iron, although an internal cement mortar lining usually serves to inhibit corrosion from the fluid being distributed, and various types of external coating are used to inhibit corrosion from the environment. Ductile iron pipe is a direct development of earlier cast iron pipe which it has superseded. Ductile iron has proven to be a better pipe material, being stronger and more fracture resistant; however, like most ferrous materials, it is susceptible to corrosion and retains some brittle characteristics. Relatively recent developments such as polyethylene sleeving and/or zinc and polymer coatings have done much to mitigate corrosion concerns, and a typical life expectancy is in excess of 75 years.

Dimensions

Ductile iron pipe is sized according to a dimensionless term known as the Pipe Size or Nominal Diameter (known by its French abbreviation, DN). This is roughly equivalent to the pipe's internal diameter in inches or millimetres. However, it is the external diameter of the pipe that is kept constant between changes in wall thickness, in order to maintain compatibility in joints and fittings. Consequently the internal diameter varies, sometimes significantly, from its nominal size. Nominal pipe sizes vary from 3 inches up to 64 inches, in increments of at least 1 inch, in the USA.

Pipe dimensions are standardised to the mutually incompatible AWWA C151 (U.S. Customary Units) in the USA, ISO 2531/EN 545/598 (metric) in Europe, and AS/NZS 2280 (metric) in Australia and New Zealand. Although both metric, European and Australian are not compatible and pipes of identical nominal diameters have quite different dimensions.

North America

Table: *Pipe dimensions according to the American AWWA C-151*

Pipe Size	*Outside Diameter [in (mm)]*
3	3.96 (100.584)
4	4.80 (121.92)
6	6.90 (175.26)
8	9.05 (229.87)
10	11.10 (281.94)
12	13.20 (335.28)
14	15.30 (388.62)
16	17.40 (441.96)
18	19.50 (495.3)
20	21.60 (548.64)
24	25.80 (655.32)
30	32.00 (812.8)

Europe

European pipe is standardized to ISO 2531 and its descendent specifications EN 545 (potable water) and EN 598 (sewage).

DN	*Outside Diameter [mm (in)]*	*Wall thickness[mm (in)]*		
		Class 40	*K9*	*K10*
40	56 (2.205)	4.8 (0.189)	6.0 (0.236)	6.0 (0.236)
50	66 (2.598)	4.8 (0.189)	6.0 (0.236)	6.0 (0.236)
60	77 (3.031)	4.8 (0.189)	6.0 (0.236)	6.0 (0.236)
65	82 (3.228)	4.8 (0.189)	6.0 (0.236)	6.0 (0.236)
80	98 (3.858)	4.8 (0.189)	6.0 (0.236)	6.0 (0.236)
100	118 (4.646)	4.8 (0.189)	6.0 (0.236)	6.0 (0.236)
125	144 (5.669)	4.8 (0.189)	6.0 (0.236)	6.0 (0.236)
150	170 (6.693)	5.0 (0.197)	6.0 (0.236)	6.5 (0.256)
200	222 (8.740)	8.4 (0.331)	6.3 (0.248)	7.0 (0.276)

Contd...

DN	*Outside Diameter [mm (in)]*	*Wall thickness[mm (in)]*		
		Class 40	*K9*	*K10*
250	274 (10.787)	5.8 (0.228)	6.8 (0.268)	7.5 (0.295)
300	326 (12.835)	6.2 (0.244)	7.2 (0.283)	8.0 (0.315)
350	378 (14.882)	7.0 (0.276)	7.7 (0.303)	8.5 (0.335)
400	429 (16.890)	7.8 (0.307)	8.1 (0.319)	9.0 (0.354)
450	480 (18.898)	-	8.6 (0.339)	9.5 (0.374)
500	532 (20.945)	-	9.0 (0.354)	10.0 (0.394)
600	635 (25.000)	-	9.9 (0.390)	11.1 (0.437)
700	738 (29.055)	-	10.9 (0.429)	12.0 (0.472)
800	842 (33.150)	-	11.7 (0.461)	13.0 (0.512)
900	945 (37.205)	-	12.9 (0.508)	14.1 (0.555)
1000	1,048 (41.260)	-	13.5 (0.531)	15.0 (0.591)
1100	1,152 (45.354)	-	14.4 (0.567)	16.0 (0.630)
1200	1,255 (49.409)	-	15.3 (0.602)	17.0 (0.669)
1400	1,462 (57.559)	-	17.1 (0.673)	19.0 (0.748)
1500	1,565 (61.614)	-	18.0 (0.709)	20.0 (0.787)
1600	1,668 (65.669)	-	18.9 (0.744)	51.0 (2.008)
1800	1,875 (73.819)	-	20.7 (0.815)	23.0 (0.906)
2000	2,082 (81.969)	-	22.5 (0.886)	25.0 (0.984)

European pipes are sized to approximately match the internal diameter of the pipe, following internal lining, to the nominal diameter. ISO 2531 maintains dimensional compatibility with older German cast iron pipes. Older British pipes, however, which used the incompatible imperial standard, BS 78, require adapter pieces when connecting to newly installed pipe. Coincidentally, the British harmonization with European pipe standards occurred at approximately the same time as its transition to ductile iron, so almost all cast iron pipe is imperial and all ductile pipe is metric.

Australia

Australian and New Zealand pipes are sized to an independent specification, AS/NZS 2280, that is not compatible with European pipes even though the same nomenclature is used. Australia adopted at an early point the imperial British cast iron pipe standard BS 78, and when this was retired on British adoption of ISO 2531, rather than similarly harmonizing with Europe, Australia opted for a 'soft' conversion from imperial units to metric, published as AS/NSZ 2280, with the physical outer diameters remaining unchanged, allowing continuity of manufacture and backwards compatibility. Therefore the inner diameters of lined pipe differ widely from the nominal diameter, and hydraulic calculations require some knowledge of the pipe standard.

Nominal Size (DN)	*Outside Diameter Size (DN)*	*Nominal Wall Thickness [mm (in)]*		*Flange Class*
		PN 20	*PN 35*	
100	122 (4.803)	-	5.0 (0.197)	7.0
150	177 (6.969)	-	5.0 (0.197)	8.0
200	232 (9.134)	-	5.0 (0.197)	8.0
225	259 (10.197)	5.0 (0.197)	5.2 (0.205)	9.0
250	286 (11.260)	5.0 (0.197)	5.6 (0.220)	9.0
300	345 (13.583)	5.0 (0.197)	6.3 (0.248)	10.0
375	426 (16.772)	5.1 (0.201)	7.3 (0.287)	10.0
450	507 (19.961)	5.6 (0.220)	8.3 (0.327)	11.0
500	560 (22.047)	6.0 (0.236)	9.0 (0.354)	12.0
600	667 (26.260)	6.8 (0.268)	310.3 (0.406)	13.0
750	826 (32.520)	7.9 (0.311)	12.2 (0.480)	15.0

Joints

Individual lengths of ductile iron pipe are joined either by flanges, couplings, or some form of spigot and socket arrangement.

Flanges

Flanges are flat rings around the end of pipes which mate with an equivalent flange from another pipe, the two being held together by bolts usually passed through holes drilled through the flanges. A deformable gasket, usually elastomeric, placed between raised faces on the mating flanges provides the seal. Flanges are designed to a large number of specifications that differ because of dimensional variations in pipes sizes and pressure requirements, and because of independent standards development.

In the U.S. flanges are either threaded or welded onto the pipe. In the European market flanges are usually welded on to the pipe. In the U.S. flanges are available in a standard 125 lb. bolt pattern as well as a 250 lb (and heavier) bolt pattern (steel bolt pattern). Both are usually rated at 250 psi (1,700 kPa). A flanged joint is rigid and can bear both tension and compression as well as a limited degree of shear and bending. It is also can be dismantled after assembly. Flanged joints cannot, however, be reliably used for buried pipe because of the possibility of soil movement placing very large bending stress on the joint.

Current flange standards used in the water industry are ANSI B16.1 in the USA, EN 1092 in Europe, and AS/NZS 4087 in Australia and New Zealand.

Spigot and Socket

Spigot and sockets involve a normal pipe end, the spigot, being inserted into the socket or bell of another pipe or fitting with a seal being made between the two within the socket. Normal spigot and socket joints do not allow direct metal to metal contact with all forces being transmitted through the elastomeric seal. They can consequently flex and allow some degree of rotation, allowing pipes to shift and relieve stresses imposed by soil movement. The corollary is that unrestrained spigot and socket joints transmit essentially no compression or tension along the axis of the pipe and little shear. Any bends, tees or valves therefore require either a restrained joint or, more commonly, thrust blocks, which transmit the forces as compression into the surrounding soil.

A large number of different socket and seals exist. The most modern is the 'push-joint' or 'slip-joint', whereby the socket and rubber seal is designed to allow the pipe spigot to be, after lubrication, simply pushed into the socket. Push joints remain proprietary designs. Also available are locking gasket systems. These locking gasket systems allow the pipe to be pushed together but do not allow the joint to come apart without using a special tool or torch on the gasket. The earliest spigot and socket cast iron pipes were jointed by filling the socket with a mixture of water, sand, iron filings and sal-ammoniac (ammonium chloride.) A gaskin ring was pushed into the socket round the spigot to contain the mixture which was pounded into the socket with a caulking tool and then pointed off. This took several weeks to set and produced a completely rigid joint. Such pipe systems are often to be seen in nineteenth century churches in the heating system.

Manufacture

Ductile iron pipe is produced by a technique known as centrifugal casting, originally developed for cast iron pipe in 1918. The molten ductile iron is poured into a rapidly spinning water-cooled mold. Centrifugal force results in an even spread of iron around the circumference.

Internal Coatings

Ductile iron pipe is somewhat resistant to internal corrosion in potable water and less aggressive forms of sewage. However, even where pipe material loss and consequently pipe wall reduction is slow, the deposition of corrosion products on the internal pipe wall can dramatically reduce the effective internal diameter and effectively choke flow, increasing pumping costs and lowering system pressure, long before the pipe itself is at risk of failure. A variety of linings are available to reduce or eliminate corrosion, including cement mortar, polyurethane and polyethylene. Of these, cement mortar lining is by far the most common.

Cement Mortar

The predominant form of lining for water applications is cement mortar centrifugally applied during manufacturing. The cement mortar comprises a mixture of cement and sand to a ratio of between 1:2 and 1:3.5. For potable water, portland cement is used; for sewage it is common to use sulfate resisting or high alumina cement.

Cement mortar linings have been found to dramatically reduce internal corrosion. A DIPRA survey has demonstrated that the Hazen-Williams factor of cement lining remains between 130 and 151 with only slight reduction with age.

External Coatings

Unprotected ductile iron, similarly to cast iron, is intrinsically resistant to corrosion in most, although not all, soils. Nonetheless, because of frequent lack of information on soil aggressiveness and to extend the installed life of buried pipe, ductile iron pipe is commonly protected by one or more external coatings. In the U.S. and Australia, loose polyethylene sleeving is preferred. In Europe, standards recommend a more sophisticated system of directly bonded zinc coatings overlaid by a finishing layer be used in conjunction with polyethylene sleeving.

Polyethylene

Polyethylene sleeving was first developed by CIPRA (since 1979, DIPRA) in the U.S. in 1951 for use in highly corrosive soil in Birmingham, Alabama. It was employed more widely in the U.S. in the late 1950s and first employed in the UK in 1965 and Australia in the mid-1960s.

Polyethylene sleeving comprises a loose sleeve of polyethylene sheet that completely wraps the pipe, including the bells of any joints. Sleeving inhibits corrosion by a number of mechanisms. It physically separates the pipe from soil particles, preventing direct galvanic corrosion. By providing an impermeable barrier to ground water, the sleeve also inhibits the diffusion of oxygen to the ductile iron surface and limits the availability of electrolytes that would accelerate corrosion. It provides a homogeneous environment along the pipe surface so that corrosion occurs evenly over the pipe. The sleeve also restricts the availability of nutrients which could support sulfate-reducing bacteria, inhibiting microbially-induced corrosion.

Sleeving is not designed to be completely water-tight but rather to greatly restrict the movement of water to and from the pipe surface. Water present beneath the sleeve and in contact with the pipe surface is rapidly deoxygenated and depleted of nutrients and forms a stable environment in which limited further corrosion occurs. An improperly installed sleeve that continues to allow the free flow of ground water is not effective in inhibiting corrosion.

Polyethylene sleeves are available in a number of materials. The most common contemporary compositions are linear low-density polyethylene film which requires an 8 mil or 200 μm thickness and high-density cross-laminated polyethylene film which requires only a 4 mil or 100 μm thickness. The latter may or may not be reinforced with a scrim layer. Polyethylene sleeving does have limitations. In European practice, its use in the absence of additional zinc and epoxy protective coatings is discouraged where natural soil resistivity is below 750 ohm/cm, where resistivity is below 1500 ohm/cm and where the pipe is installed at or below the water table, where there are additional artificial soil contaminants or where there are stray currents. Because of the vulnerability of polyethylene to UV degradation, sleeving, or sleeved pipe should not be stored in sunlight, although carbon pigments included in the sleeving can provide some limited protection. Polyethylene sleeving is standardised according to ISO 8180 internationally, AWWA C105 in the U.S., BS 6076 in the UK and AS 3680 and AS 3681 in Australia.

Zinc

In Europe, ductile iron pipe is typically manufactured with a zinc coating overlaid by either a bituminous, polymer, or epoxy finishing layer. EN 545/598 mandates a minimum zinc content of 200 g/m^2 (with local minima of 110 g/m^2 at 99.99% purity) and a minimum average finishing layer thickness of 70 μm (with local minima of 50 μm).

No current AWWA standards are available for bonded coatings (zinc, coal tar epoxy, tape-wrap systems as seen on steel pipe) for ductile iron pipe, DIPRA does not endorse bonded coatings, and AWWA M41 generally views them unfavourably, recommending they be used only in conjunction with cathodic protection.

Bituminous Coatings

Zinc coatings are generally not employed in the U.S. and Australia. In order to protect ductile iron pipe prior to installation, pipe is instead supplied with a temporary 1 mil or 25 μm thick bituminous coating. This coating is not intended to provide protection once the pipe is installed.

Industry Associations

In the United States ductile iron pipe is often promoted to municipalities and consulting engineers by the Ductile Iron Pipe Research Association. Their focus is to promote the benefits of using ductile iron pipe on utility projects (water & sewer) over alternate products such as PVC, PCP, and HDPE.

Environmental

Ductile iron pipe in the developed world is normally manufactured exclusively from scrap steel. Ductile iron pipe can be recycled. In the U.S. with the growing environmental movement ductile iron pipe is in a natural position to regain market share lost to its largest competitor, the PVC industry, over the past 40 years.

Hollow Structural Section

A hollow structural section (HSS) is a type of metal profile with a hollow tubular cross section. In some countries they are referred to instead as a structural hollow section (SHS).

Most HSS are of circular or rectangular section, although other shapes are available, such as elliptical. HSS is only composed of structural steel per code.

HSS is sometimes mistakenly referenced as hollow structural steel. Rectangular HSS are also called tube steel or structural tubing. Circular HSS are sometimes mistakenly called steel pipe though true steel pipe is actually dimensioned and classed differently than HSS. The corners of HSS are heavily rounded, or chamfered, at radii approximately twice the wall thickness. The wall thickness is uniform around the section.

In the UK, the terms are circular and rectangular hollow section (CHS and RHS). However, the dimensions and tolerances differ slightly from HSS.

Use in Structures

HSS, especially rectangular sections, are commonly used in welded steel frames where members experience loading in multiple directions.

Square and circular HSS have very efficient shapes for this multiple-axis loading as they have uniform geometric and thus uniform strength characteristics along two or more cross-sectional axes; this makes them good choices for columns. They also have excellent resistance to torsion.

HSS can also be used as beams, although wide flange or I-beam shapes are in many cases a more efficient structural shape for this application. However, the HSS has superior resistance to lateral torsional buckling.

The flat square surfaces of rectangular HSS can ease construction, and they are sometimes preferred for architectural aesthetics in exposed structures, although elliptical HSS are becoming more popular in exposed structures for the same aesthetic reasons.

HSS is commonly available in mild steel, such as A500 grade B.

Manufacture

Square HSS is made the same way as pipe. During the manufacturing process flat steel plate is gradually changed in shape to become round where the edges are presented ready to weld. The edges are then welded together to form the mother tube. During the manufacturing process the mother tube goes through a series of shaping stands which form the round HSS (mother tube) into the final square or rectangular shape. Most American manufacturers adhere to the ASTM A500 standard, while Canadian manufacturers follow both ASTM A500 and CSA G40.21. European hollow sections are generally in accordance with the EN 10210 standard. IT is one or the structural steel element in a construction.

Filled HSS

HSS is often filled with concrete to improve fire rating, as well as robustness. When this is done, the product is referred to as a lolly column. Lally column is the preferred spelling however, since it got its name from the Lally Column Company. For example, barriers around parking areas, bollards, made of HSS are often filled, to at least bumper height, with concrete. This is an inexpensive (when replacement costs are factored in) way of adding compressive strength to the bollard, which can help prevent unsightly local denting, though does not generally significantly increase the overall structural properties of the bollard.

Hydraulic Pipes

Hydraulic machines are machinery and tools that use liquid fluid power to do simple work. Heavy equipment is a common example. In this type of machine, hydraulic fluid is transmitted throughout the machine to various hydraulic motors and hydraulic cylinders and which becomes pressurised according to the resistance present. The fluid is controlled directly or automatically by control valves and distributed through hoses and tubes.

The popularity of hydraulic machinery is due to the very large amount of power that can be transferred through small tubes and flexible hoses, and the high power density and wide array of actuators that can make use of this power.

Hydraulic machinery is operated by the use of hydraulics, where a liquid is the powering medium.

Force and Torque Multiplication

A fundamental feature of hydraulic systems is the ability to apply force or torque multiplication in an easy way, independent of the distance between the input and output, without the need for mechanical gears or levers, either by altering the effective areas in two connected cylinders or the effective displacement (cc/rev) between a pump and motor. In normal cases, hydraulic ratios are combined with a mechanical force or torque ratio for optimum machine designs such as boom movements and trackdrives for an excavator.

Examples

Cylinder C1 is one inch in radius, and cylinder C2 is ten inches in radius. If the force exerted on C1 is 10 lbf, the force exerted by C2 is 1000 lbf because C2 is a hundred times larger in area ($S = \pi r^2$) as C1. The downside to this is that you have to move C1 a hundred inches to move C2 one inch. The most common use for this is the classical hydraulic jack where a pumping cylinder with a small diameter is connected to the lifting cylinder with a large diameter.

Pump and Motor

If a hydraulic rotary pump with the displacement 10 cc/rev is connected to a hydraulic rotary motor with 100 cc/rev, the shaft torque required to drive the pump is 10 times less than the torque available at the motor shaft, but the shaft speed (rev/min) for the motor is 10 times less than the pump shaft speed. This combination is actually the same type of force multiplication as the cylinder example (1) just that the linear force in this case is a rotary force, defined as torque.

Both these examples are usually referred to as a hydraulic transmission or hydrostatic transmission involving a certain hydraulic "gear ratio".

Hydraulic Circuits

For the hydraulic fluid to do work, it must flow to the actuator and or motors, then return to a reservoir. The fluid is then filtered and re-pumped. The path taken by hydraulic fluid is called a hydraulic circuit of which there are several types. Open center circuits use pumps which supply a continuous flow. The flow is returned to *tank* through the control valve's *open center*; that is, when the control valve is centred, it provides an open return path to tank and the fluid is not pumped to a high pressure. Otherwise, if the control valve is actuated it routes fluid to and from an actuator and tank. The fluid's pressure will rise to meet any resistance, since the pump has a constant output. If the pressure rises too high, fluid returns to tank through a pressure relief valve. Multiple control valves may be stacked in series. This type of circuit can use inexpensive, constant displacement pumps.

Closed center circuits supply full pressure to the control valves, whether any valves are actuated or not. The pumps vary their flow rate, pumping very little hydraulic fluid until the operator actuates a valve. The valve's spool therefore doesn't need an open center return path to tank. Multiple valves can be connected in a parallel arrangement and system pressure is equal for all valves.

Constant Pressure and Load-Eensing Systems

The closed center circuits exist in two basic configurations, normally related to the regulator for the variable pump that supplies the oil:

Constant pressure systems (CP-system), standard. Pump pressure always equals the pressure setting for the pump regulator. This setting must cover the maximum required load pressure. Pump delivers flow according to required sum of flow to the consumers. The CP-system generates large power losses if the machine works with large variations in load pressure and the average system pressure is much lower than the pressure setting for the pump regulator. CP is simple in design. Works like a pneumatic system. New hydraulic functions can easily be added and the system is quick in response.

Constant pressure systems (CP-system), unloaded. Same basic configuration as 'standard' CP-system but the pump is unloaded to a low stand-by pressure when all valves are in neutral position. Not so fast response as standard CP but pump lifetime is prolonged.

Load-sensing systems (LS-system) generates less power losses as the pump can reduce both flow and pressure to match the load requirements, but requires more tuning than the CP-system with respect to system stability. The LS-system also requires additional logical valves and compensator valves in the directional valves, thus it is technically more complex and more expensive than the CP-system. The LS-system system generates a constant power loss related to the regulating pressure drop for the pump regulator:

$$Powerloss = \Delta p_{LS} \cdot Q_{tot}$$

The average Δp_{LS} is around 2 MPa (290 psi). If the pump flow is high the extra loss can be considerable. The power loss also increase if the load pressures varies a lot. The cylinder areas, motor displacements and mechanical torque arms must be designed to match in load pressure in order to bring down the power losses. Pump pressure always equals the maximum load pressure when several functions are run simultaneously and the power input to the pump equals the (max. load pressure + Δp_{LS}) x sum of flow.

Five Basic Types of Load-Sensing Systems

(1) Load sensing *without compensators* in the directional valves. Hydraulically controlled LS-pump.

(2) Load sensing *with up-stream compensator* for each connected directional valve. Hydraulically controlled LS-pump.

(3) Load sensing *with down-stream compensator* for each connected directional valve. Hydraulically controlled LS-pump.

(4) Load sensing *with a combination of up-stream and down-stream compensators*. Hydraulically controlled LS-pump.

(5) Load sensing with synchronized, both electric controlled pump displacement and electric controlled valve flow area for faster response, increased stability and less system losses. This is a new type of LS-system, not yet fully developed.

Technically the down-stream mounted compensator in a valveblock can physically be mounted "up-stream", but work as a down-stream compensator. System type (3) gives the advantage that activated functions are synchronized independent of pump flow capacity. The flow relation between 2 or more activated functions remains independent of load pressures even if the pump reach the maximum swivel angle. This feature is important for machines that often run with the pump at maximum swivel angel and with several activated functions that must be synchronized in speed, such as with excavators. With type (4) system, the functions with *up-stream* compensators have priority. Example: Steering-function for a wheel loader. The system type with down-stream compensators usually have a unique trademark depending on the manufacturer of the valves, for example “LSC” (Linde Hydraulics), “LUDV” (Bosch Rexroth Hydraulics) and “Flowsharing” (Parker Hydraulics) etc. No official standardized name for this type of system has been established but Flowsharing is a common name for it.

Open and Closed Circuits

Open-loop: Pump-inlet and motor-return (via the directional valve) are connected to the hydraulic tank. The term loop applies to feedback; the more correct term is open versus closed “circuit”. Closed-loop: Motor-return is connected directly to the pump-inlet.

To keep up pressure on the low pressure side, the circuits have a charge pump (a small gearpump) that supplies cooled and filtered oil to the low pressure side. Closed-loop circuits are generally used for hydrostatic transmissions in mobile applications. *Advantages:* No directional valve and better response, the circuit can work with higher pressure. The pump swivel angle covers both positive and negative flow direction. *Disadvantages:* The pump cannot be utilized for any other hydraulic function in an easy way and cooling can be a problem due to limited exchange of oil flow. High power closed loop systems generally must have a ‘flush-valve’ assembled in the circuit in order to exchange much more flow than the basic leakage flow from the pump and the motor, for increased cooling and filtering. The flush valve is normally integrated in the motor housing to get a cooling effect for the oil that is rotating in the motor housing itself. The losses in the motor housing from rotating effects and losses in the ball bearings can be considerable as motor speeds will reach 4000-5000 rev/min or even more at maximum vehicle speed. The leakage flow as well as the extra flush flow must be supplied by the charge pump. Large charge pumps is thus very important if the transmission is designed for high pressures and high motor speeds. High oil temperatures, is usually a major problem when using hydrostatic transmissions at high vehicle speeds for longer periods, for instance when transporting the machine from one work place to the other. High oil temperatures for long periods will drastically reduce the lifetime for the transmission. To keep down the oil temperature, the system pressure during transport must be lowered, meaning that the minimum displacement for the motor must be limited to a reasonable value. Circuit pressures during transport around 200-250 bar is recommended.

Closed loop systems in mobile equipment are generally used for the transmission as an alternative to mechanical and hydrodynamic (converter) transmissions. The advantage is a stepless gear ratio (continuously variable speed/torque) and a more flexible control of the gear ratio depending on the load and operating conditions. The hydrostatic transmission is generally limited to around 200 kW maximum power, as the total cost gets too high at higher power compared to a hydrodynamic transmission. Large wheel loaders for instance and heavy machines are therefore usually equipped with converter transmissions. Recent technical achievements for the converter transmissions have improved the efficiency and developments in the software have also improved the characteristics, for example selectable gear shifting προγραμμες during operation and more gear steps, giving them characteristics close to the hydrostatic transmission.

Hydrostatic transmissions for earth moving machines, such as for track loaders, are often equipped with a separate 'inch pedal' that is used to temporarily increase the diesel engine rpm while reducing the vehicle speed in order to increase the available hydraulic power output for the working hydraulics at low speeds and increase the tractive effort. The function is similar to stalling a converter gearbox at high engine rpm. The inch function affects the preset characteristics for the 'hydrostatic' gear ratio versus diesel engine rpm.

Components—Hydraulic Pump

Hydraulic pumps supply fluid to the components in the system. Pressure in the system develops in reaction to the load. Hence, a pump rated for 5,000 psi is capable of maintaining flow against a load of 5,000 psi. Pumps have a power density about ten times greater than an electric motor (by volume). They are powered by an electric motor or an engine, connected through gears, belts, or a flexible elastomeric coupling to reduce vibration.

Common types of hydraulic pumps to hydraulic machinery applications are;

- Gear pump: cheap, durable, simple. Less efficient, because they are constant (fixed) displacement, and mainly suitable for pressures below 20 MPa (3000 psi).
- Vane pump: cheap and simple, reliable (especially in g-rotor form). Good for higher-flow low-pressure output.
- Axial piston pump: many designed with a variable displacement mechanism, to vary output flow for automatic control of pressure. There are various axial piston pump designs, including swashplate (sometimes referred to as a valveplate pump) and checkball (sometimes referred to as a wobble plate pump). The most common is the swashplate pump. A variable-angle swashplate causes the pistons to reciprocate a greater or lesser

distance per rotation, allowing output flow rate and pressure to be varied (greater displacement angle causes higher flow rate, lower pressure, and vice versa).

- Radial piston pump A pump that is normally used for very high pressure at small flows.

Piston pumps are more expensive than gear or vane pumps, but provide longer life operating at higher pressure, with difficult fluids and longer continuous duty cycles. Piston pumps make up one half of a hydrostatic transmission.

Control Valves

Directional control valves route the fluid to the desired actuator. They usually consist of a spool inside a cast iron or steel housing. The spool slides to different positions in the housing, intersecting grooves and channels route the fluid based on the spool's position.

The spool has a central (neutral) position maintained with springs; in this position the supply fluid is blocked, or returned to tank. Sliding the spool to one side routes the hydraulic fluid to an actuator and provides a return path from the actuator to tank. When the spool is moved to the opposite direction the supply and return paths are switched. When the spool is allowed to return to neutral (center) position the actuator fluid paths are blocked, locking it in position.

Directional control valves are usually designed to be stackable, with one valve for each hydraulic cylinder, and one fluid input supplying all the valves in the stack. Tolerances are very tight in order to handle the high pressure and avoid leaking, spools typically have a clearance with the housing of less than a thousandth of an inch (25 μm). The valve block will be mounted to the machine's frame with a *three point* pattern to avoid distorting the valve block and jamming the valve's sensitive components. The spool position may be actuated by mechanical levers, hydraulic *pilot* pressure, or solenoids which push the spool left or right. A seal allows part of the spool to protrude outside the housing, where it is accessible to the actuator.

The main valve block is usually a stack of *off the shelf* directional control valves chosen by flow capacity and performance. Some valves are designed to be proportional (flow rate proportional to valve position), while others may be simply on-off. The control valve is one of the most expensive and sensitive parts of a hydraulic circuit.

- Pressure relief valves are used in several places in hydraulic machinery; on the return circuit to maintain a small amount of pressure for brakes, pilot lines, etc... On hydraulic cylinders, to prevent overloading and hydraulic line/seal rupture. On the hydraulic reservoir, to maintain a small positive pressure which excludes moisture and contamination.

- Pressure regulators reduce the supply pressure of hydraulic fluids as needed for various circuits.
- Sequence valves control the sequence of hydraulic circuits; to ensure that one hydraulic cylinder is fully extended before another starts its stroke, for example.
- Shuttle valves provide a logical or function.
- Check valves are one-way valves, allowing an accumulator to charge and maintain its pressure after the machine is turned off, for example.
- Pilot controlled Check valves are one-way valve that can be opened (for both directions) by a foreign pressure signal. For instance if the load should not be held by the check valve anymore. Often the foreign pressure comes from the other pipe that is connected to the motor or cylinder.
- Counterbalance valves are in fact a special type of pilot controlled check valve. Whereas the check valve is open or closed, the counterbalance valve acts a bit like a pilot controlled flow control.
- Cartridge valves are in fact the inner part of a check valve; they are *off the shelf* components with a standardized envelope, making them easy to populate a proprietary valve block. They are available in many configurations; on/off, proportional, pressure relief, etc. They generally screw into a valve block and are electrically controlled to provide logic and automated functions.
- Hydraulic fuses are in-line safety devices designed to automatically seal off a hydraulic line if pressure becomes too low, or safely vent fluid if pressure becomes too high.
- Auxiliary valves in complex hydraulic systems may have auxiliary valve blocks to handle various duties unseen to the operator, such as accumulator charging, cooling fan operation, air conditioning power, etc. They are usually custom valves designed for the particular machine, and may consist of a metal block with ports and channels drilled. Cartridge valves are threaded into the ports and may be electrically controlled by switches or a microprocessor to route fluid power as needed.

Actuators

- Hydraulic cylinder
- Swashplates are used in 'hydraulic motors' requiring highly accurate control and also in 'no stop' continuous (360°) precision positioning mechanisms. These are frequently driven by several hydraulic pistons acting in sequence.
- Hydraulic motor (a pump plumbed in reverse)
- hydrostatic transmission
- Brakes.

Reservoir

The hydraulic fluid reservoir holds excess hydraulic fluid to accommodate volume changes from: cylinder extension and contraction, temperature driven expansion and contraction, and leaks. The reservoir is also designed to aid in separation of air from the fluid and also work as a heat accumulator to cover losses in the system when peak power is used. Design engineers are always pressured to reduce the size of hydraulic reservoirs, while equipment operators always appreciate larger reservoirs. Reservoirs can also help separate dirt and other particulate from the oil, as the particulate will generally settle to the bottom of the tank.

Some designs include dynamic flow channels on the fluid's return path that allow for a smaller reservoir.

Accumulators

Accumulators are a common part of hydraulic machinery. Their function is to store energy by using pressurized gas. One type is a tube with a floating piston. On one side of the piston is a charge of pressurized gas, and on the other side is the fluid. Bladders are used in other designs. Reservoirs store a system's fluid.

Examples of accumulator uses are backup power for steering or brakes, or to act as a shock absorber for the hydraulic circuit.

Hydraulic Fluid

Also known as *tractor fluid*, hydraulic fluid is the life of the hydraulic circuit. It is usually petroleum oil with various additives. Some hydraulic machines require fire resistant fluids, depending on their applications. In some factories where food is prepared, either an edible oil or water is used as a working fluid for health and safety reasons.

In addition to transferring energy, hydraulic fluid needs to lubricate components, suspend contaminants and metal filings for transport to the filter, and to function well to several hundred degrees Fahrenheit or Celsius.

Filters

Filters are an important part of hydraulic systems. Metal particles are continually produced by mechanical components and need to be removed along with other contaminants.

Filters may be positioned in many locations. The filter may be located between the reservoir and the pump intake. Blockage of the filter will cause cavitation and possibly failure of the pump. Sometimes the filter is located between the pump and the control valves. This arrangement is more expensive, since the filter housing is pressurized, but eliminates cavitation problems and protects the control valve from pump failures. The third common filter location is just before the return line enters the reservoir. This location is relatively insensitive to blockage and does not require a pressurized housing, but contaminants that enter the reservoir from external sources are not filtered until passing through the system at least once.

Tubes, Pipes and Hoses

Hydraulic tubes are seamless steel precision pipes, specially manufactured for hydraulics. The tubes have standard sizes for different pressure ranges, with standard diameters up to 100 mm. The tubes are supplied by manufacturers in lengths of 6 m, cleaned, oiled and plugged. The tubes are interconnected by different types of flanges (especially for the larger sizes and pressures), welding cones/nipples (with o-ring seal), several types of flare connection and by cut-rings. In larger sizes, hydraulic pipes are used. Direct joining of tubes by welding is not acceptable since the interior cannot be inspected.

Hydraulic pipe is used in case standard hydraulic tubes are not available. Generally these are used for low pressure. They can be connected by threaded connections, but usually by welds. Because of the larger diameters the pipe can usually be inspected internally after welding. Black pipe is non-galvanized and suitable for welding.

Hydraulic hose is graded by pressure, temperature, and fluid compatibility. Hoses are used when pipes or tubes can not be used, usually to provide flexibility for machine operation or maintenance. The hose is built up with rubber and steel layers. A rubber interior is surrounded by multiple layers of woven wire and rubber. The exterior is designed for abrasion resistance. The bend radius of hydraulic hose is carefully designed into the machine, since hose failures can be deadly, and violating the hose's minimum bend radius will cause failure. Hydraulic hoses generally have steel fittings swaged on the ends. The weakest part of the high pressure hose is the connection of the hose to the fitting. Another disadvantage of hoses is the shorter life of rubber which requires periodic replacement, usually at five to seven year intervals.

Tubes and pipes for hydraulic applications are internally oiled before the system is commissioned. Usually steel piping is painted outside. Where flare and other couplings are used, the paint is removed under the nut, and is a location where corrosion can begin. For this reason, in marine applications most piping is stainless steel.

Seals, Fittings and Connections

In general, valves, cylinders and pumps have female threaded bosses for the fluid connection, and hoses have female ends with captive nuts. A male-male fitting is chosen to connect the two. Many standardized systems are in use.

Fittings serve several purposes;

1. To bridge different standards; O-ring boss to JIC, or pipe threads to face seal, for example.
2. To allow proper orientation of components, a 90°, 45°, straight, or swivel fitting is chosen as needed. They are designed to be positioned in the correct orientation and then tightened.
3. To incorporate bulkhead hardware.
4. A *quick disconnect* fitting may be added to a machine without modification of hoses or valves.

A typical piece of heavy equipment may have thousands of sealed connection points and several different types:

- Pipe fittings, the fitting is screwed in until tight, difficult to orient an angled fitting correctly without over or under tightening.
- O-ring boss, the fitting is screwed into a boss and orientated as needed, an additional nut tightens the fitting, washer and o-ring in place.
- Flare fittings, are metal to metal compression seals deformed with a cone nut and pressed into a flare mating.
- Face seal, metal flanges with a groove and o-ring are fastened together.

- Beam seals are costly metal to metal seals used primarily in aircraft.
- Swaged seals, tubes are connected with fittings that are swaged permanently in place. Primarily used in aircraft.

Elastomeric seals (O-ring boss and face seal) are the most common types of seals in heavy equipment and are capable of reliably sealing 6000+ psi (40+ MPa) of fluid pressure.

Basic Calculations

Hydraulic power is defined as flow times pressure. The hydraulic power supplied by a pump:

Power = (P x Q) ÷ 600

where power is in kilowatts [kW], P pressure in bars, and Q is the flow in litres per minute. For example, a pump delivers 180 lit/min and the pressure equals 250 bar, therefore the power of the pump is 75 kW.

When calculating the power input to the pump, the total pump efficiency η_{total} must be included. This efficiency is the product of volumetric efficiency, η_{vol} and the hydromechanical efficiency, η_{hm}. Power input = Power output ÷ η_{total}.

The average for axial piston pumps, $\eta_{total} = 0.87$. In the example the power source, for example a diesel engine or an electric motor, must be capable of delivering at least 75 ÷ 0.87 = 86 [kW]. The hydraulic motors and cylinders that the pump supplies with hydraulic power also have efficiencies and the total system efficiency (without including the pressure drop in the hydraulic pipes and valves) will end up at approx. 0.75.

Cylinders normally have a total efficiency around 0.95 while hydraulic axial piston motors 0.87, the same as the pump. In general the power loss in a hydraulic energy transmission is thus around 25% or more at ideal viscosity range 25-35 [cSt].

Calculation of the required max. power output for the diesel engine, rough estimation:

(1) Check the max. powerpoint, i.e. the point where pressure times flow reach the max. value.

(2) $E_{diesel} = (P_{max} \cdot Q_{tot}) \div \eta$.

Q_{tot} = calculate with the theoretical pump flow for the consumers not including leakages at max. power point.

P_{max} = actual pump pressure at max. power point.

Note: η is the total efficiency = (output mechanical power ÷ input mechanical power). For rough estimations, η = 0.75. Add 10-20% (depends on the application) to this power value.

(3) Calculate the required pumpdisplacement from required max. sum of flow for the consumers in worst case and the diesel engine rpm in this point. The max. flow can differ from the flow used for calculation of the diesel engine power. Pump volumetric efficiency average, piston pumps: η_{vol}= 0.93.

Pumpdisplacement $V_{pump} = Q_{tot} \div n_{diesel} \div 0.93$.

(4) Calculation of prel. cooler capacity: Heat dissipation from hydraulic oil tanks, valves, pipes and hydraulic components is less than a few percent in standard mobile equipment and the cooler capacity must include some margins. Minimum cooler capacity, $E_{cooler} = 0.25E_{diesel}$

At least 25% of the input power must be dissipated by the cooler when peak power is utilized for long periods.

In normal case however, the peak power is used for only short periods, thus the actual cooler capacity required might be considerably less. The oil volume in the hydraulic tank is also acting as a heat accumulator when peak power is used.

The system efficiency is very much dependent on the type of hydraulic work tool equipment, the hydraulic pumps and motors used and power input to the hydraulics may vary a lot. Each circuit must be evaluated and the load cycle estimated. New or modified systems must always be tested in practical work, covering all possible load cycles. An easy way of measuring the actual average power loss in the system is to equip the machine with a test cooler and measure the oil temperature at cooler inlet, oil temperature at cooler outlet and the oil flow through the cooler, when the machine is in normal operating mode. From these figures the test cooler power dissipation can be calculated and this is equal to the power loss when temperatures are stabilized. From this test the actual required cooler can be calculated to reach specified oil temperature in the oil tank. One problem can be to assemble the measuring equipment inline, especially the oil flow meter.

Pipe and Tube Bender

Tube bending is the umbrella term for metal forming processes used to permanently form pipes or tubing. One has to differentiate between form-bound and freeform-bending procedures, as well as between heat supported and cold forming procedures.

Form bound bending procedures like "press bending" or "rotary draw bending" are used to form the work piece into the shape of a die. Straight tube stock can be formed using a bending machine to create a variety of single or multiple bends and to shape the piece into the desired form. This processes can be used to form complex shapes out of different types of ductile metal tubing. Freeform-bending processes, like *three-roll-pushbending*, shape the workpiece kinematically, thus the bending contour is not dependent on the tool geometry.

Generally, round stock is what is used in tube bending. However, square and rectangular tubes and pipes may also be bent to meet job specifications. Other factors involved in the tube bending process is the wall thickness, tooling and lubricants needed by the pipe and tube bender to best shape the material.

Geometry

A tube can be bent in multiple directions and angles. Common simple bends consist of forming *elbows*, which are bends that range from 2 to 90°, and *U-bends*, which are 180° bends. More complex geometries include multiple two-dimensional (2D) bends and three-dimensional (3D) bends. A 2D tube has the openings on the same plane; a 3D has openings on different planes.

One side effect of bending the workpiece is the wall thickness changes; the wall along the inner radius of the tube becomes thicker and the outer wall becomes thinner. To overcome this the tube may be supported internally or externally to preserve the cross section. Depending on the bend angle, wall thickness, and bending process the inside of the wall may wrinkle.

Processes

Tube bending as a process starts with loading a tube into a pipe bender and clamping it into place between two dies, the clamping block and the forming die. The tube is also loosely held by two other dies, the wiper die and the pressure die. The process of tube bending involves using mechanical force to push stock material pipe or tubing against a die, forcing the pipe or tube to conform to the shape of the die. Often, stock tubing is held firmly in place while the end is rotated and rolled around the die. Other forms of processing including pushing stock through rollers that bend it into a simple curve. For some tube bending processing, a mandrel is placed inside the tube to prevent collapsing. The tube is also held in tension by a wiper die to prevent any creasing during stress. A wiper die is usually made of a softer alloy i.e. aluminium, brass to avoid scratching or damaging the material being bent.

Much of the tooling is made of hardened steel or tooled steel to maintain and prolong the tools life. However wherever there is a concern of scratching or gouging the work piece, a softer material such as aluminium or bronze is utilized. For example, the clamping block, rotating form block and pressure die are often formed from the hardened steel because the tubing is not moving past these parts of the machine. On the other hand, the pressure die and the wiping die are formed from aluminium or bronze to maintain the shape and surface of the workpiece as it slides by.

Pipe bending machines are typically human powered, pneumatic powered, hydraulic assisted, hydraulic driven, or electric servomotor.

Press Bending

Probably the first bending process used on cold pipes and tubing. In this process a die in the shape of the bend is pressed against the pipe forcing the pipe to fit the shape of the bend. Because the pipe is not supported internally there is some deformation of the shape of the pipe giving an ovular cross section. This process is used where a consistent cross section of the pipe is not required. Although a single die can produce various shapes, it only works for one size tube and radius.

Rotary Draw Bending

Rotary draw bending (RDB) are precise in that they bend using tooling or "die sets" which have a constant center line radius (CLR). The die set consists of two parts: The *former die* creates the shape to which the material will be bent. The *counter die* does the work of pushing the material into the former die while travelling the length of the bend. Rotary draw benders can be προγραμμεmable to store multiple bend jobs with varying degrees of bending. Often a positioning index table (IDX) is attached to the bender allowing the operator to reproduce complex bends which can have multiple bends and differing planes.

Rotary draw benders are the most popular machines for use in bending tube, pipe and solids for applications like: handrails, frames, motor vehicle roll cages, handles, lines and much more. Rotary draw benders create aesthetically pleasing bends when the right tooling is matched to the application.

Three-Roll Pushbending

The *Three-Roll Pushbending* (TRPB) is the most commonly used freeform-bending process to manufacture bending geometries consisting of several plain bending curves. Nevertheless, a 3D-shaping is possible. The profile is guided between bending-roll and supporting-roll(s), while being pushed through the tools. The position of the forming-roll defines the bending radius. The bending point is the tangent-point between tube and bending-roll. To change the bending plane, the pusher rotates the tube around its longitudinal axis. Generally, a TRPB tool kit can be applied on a conventional rotary draw bending machine.

Bending contours defined as spline-or polynomial-functions can be manufactured.

Heat-Induction

An induction coil is placed around a small section of the pipe at the bend point. It is then heated to between 800 and 2,200 degrees Fahrenheit (430 and 1,200 °C). While the pipe is hot, pressure is placed on the pipe to bend it. The pipe is then quenched with either air or water spray. Heat-Induction bending is used on large pipes such as freeway signs, power plants, and petroleum pipe lines.

Roll Benders

During the roll bending process the pipe, extrusion, or solid is passed through a series of rollers (typically 3) that apply pressure to the pipe gradually changing the bend radius in the pipe. The pyramid style roll benders have one moving roll, usually the top roll. Double pinch type roll benders have two adjustable rolls, usually the bottom rolls, and a fixed top roll. This method of bending causes very little deformation in the cross section of the pipe. This process is suited to producing coils of pipe as well as long gentle bends like those used in truss systems.

Sand-Packing/Hot-Slab Forming

In the sand packing process the pipe is filled with fine sand and the ends are capped. The pipe is then heated in a furnace to 1,600 °F (870 °C) or higher. The pipe is then placed on a slab with pins set in it. The pipe is then bent around the pins using a winch, crane, or some other mechanical force. The sand in the pipe minimizes distortion in the pipe cross section.

Mandrels

A mandrel is a steel rod or linked ball inserted into the tube while it is being bent to give the tube extra support to reduce wrinkling and breaking the tube during this process. The different types of mandrels are as follows.

1. Plug mandrel, a solid rod used on normal bends.
2. Form mandrel, a solid rod with curved end used on bend when more support is need.

3. Ball mandrel without cable, unlinked steel ball bearings inserted into tube, used on critical and precise bends.
4. Ball mandrel with cable, linked ball bearings inserted into tube, used on critical bend and precise bends.
5. Sand, sand packed into tube.

In production of a product where the bend is not critical a plug mandrel can be used. A form type tapers the end of the mandrel to provide more support in the bend of the tube. When precise bending is needed a ball mandrel (or ball mandrel with steel cable) should be used. The conjoined ball-like disks are inserted into the tubing to allow for bending while maintaining the same diameter throughout. Other styles include using sand, cerrobend, or frozen water. These allow for a somewhat constant diameter while providing an inexpensive alternative to the aforementioned styles.

Performance automotive or motorcycle exhaust pipe is a common application for a mandrel.

Bending Springs

These are strong but flexible springs inserted into a pipe to support the pipe walls during manual bending. They have diameters only slightly less than the internal diameter of the pipe to be bent. They are only suitable for bending 15-and-22 mm (0.6-and-0.9 in) soft copper pipe (typically used in household plumbing).

The spring is pushed into the pipe until its center is roughly where the bend is to be. A length of flexible wire can be attached to the end of the spring to facilitate its removal. The pipe is generally held against the flexed knee, and the ends of the pipe are pulled up to create the bend. To make it easier to retrieve the spring from the pipe, it is a good idea to bend the pipe slightly more than required, and then slacken it off a little. They are less cumbersome than rotary benders, but are not suitable for bending short lengths of piping when it is difficult to get the required leverage on the pipe ends.

Bending springs for smaller diameter pipes (10 mm copper pipe) slide over the pipe instead of inside.

Pipeline Transport

Pipeline transport is the transportation of goods through a pipe. Most commonly, liquids and gases are sent, but pneumatic tubes that transport solid capsules using compressed air are also used.

As for gases and liquids, any chemically stable substance can be sent through a pipeline. Therefore sewage, slurry, water, or even beer pipelines exist; but arguably the most valuable are those transporting fuels: oil (oleoduct), natural gas (gas grid), and biofuels.

Dmitri Mendeleev first suggested using a pipe for transporting petroleum in 1863.

Types by Transported Substance

For Oil or Uatural Gas

There is some argument as to when the first crude oil pipeline was built. However, some say pipeline transport was pioneered by Vladimir Shukhov and the Branobel company in the late 19th century. Others say oil pipelines originated when the Oil Transport Association first constructed a 2-inch (51 mm) wrought iron pipeline over a 6-mile (9.7 km) track from an oil field in Pennsylvania to a railroad station in Oil Creek, in the 1860s. Pipelines are generally the most economical way to transport large quantities of oil, refined oil products or natural gas over land. Compared to shipping by railroad, they have lower cost per unit and higher capacity. Although pipelines can be built under the sea, that process is economically and technically demanding, so the majority of oil at sea is transported by tanker ships.

Oil pipelines are made from steel or plastic tubes with inner diameter typically from 4 to 48 inches (100 to 1,200 mm). Most pipelines are buried at a typical depth of about 3 to 6 feet (0.91 to 1.8 m). The oil is kept in motion by pump stations along the pipeline, and usually flows at speed of about 1 to 6 metres per second (3.3 to 20 ft/s). Multi-product pipelines are used to transport two or more different products in sequence in the same pipeline. Usually in multi-product pipelines there is no physical separation between the different products. Some mixing of adjacent products occurs, producing interface. At the receiving facilities this interface is usually absorbed in one of the product based on pre-calculated absorption rates.

Crude oil contains varying amounts of wax, or paraffin, and in colder climates wax buildup may occur within a pipeline. Often these pipelines are inspected and cleaned using pipeline inspection gauges, *pigs*, also known as *scrapers* or *Go-devils*. Smart pigs (also known as intelligent or intelligence pigs) are used to detect anomalies in the pipe such as dents, metal loss caused by corrosion, cracking or other mechanical damage. These devices are launched from pig-launcher stations and travel through the pipeline to be received at any other station down-stream, either cleaning wax deposits and material that may have accumulated inside the line or inspecting and recording the condition of the line. For natural gas, pipelines are constructed of carbon steel and varying in size from 2 to 60 inches (51 to 1,500 mm) in diameter, depending on the type of pipeline. The gas is pressurized by compressor stations and is odorless unless mixed with a mercaptan odorant where required by a regulating authority.

For Ammonia

Highly-toxic ammonia is technically the most dangerous substance to be transported through long-distance pipelines. However, practically incidents on ammonia-transporting lines are uncommon-unlike on industrial ammonia-processing equipment.

A major example of ammonia pipeline is the Ukrainian *Transammiak* line connecting the TogliattiAzot facility in Russia to the exporting Black Sea-port of Odessa.

For Biofuels (Ethanol and Biobutanol)

Pipelines have been used for transportation of ethanol in Brazil, and there are several ethanol pipeline projects in Brazil and the United States. Main problems related to the shipment of ethanol by pipeline are its high oxygen content, which makes it corrosive, and absorption of water and impurities in pipelines, which is not a problem with oil and natural gas. Insufficient volumes and cost-effectiveness are other considerations limiting construction of ethanol pipelines.

For Coal and Ore

Slurry pipelines are sometimes used to transport coal or ore from mines. The material to be transported is closely mixed with water before being introduced to the pipeline; at the far end, the material must be dried. One example is a 525 km slurry pipeline which is planned to transport iron ore from the Minas-Rio mine (producing 26.5 million tonnes per year) to a port at Aηu in Brazil. An existing example is the 85 km Savage River Slurry pipeline in Tasmania, Australia, possibly the world's first when it was built in 1967. It includes a 366m bridge span at 167m above the Savage River.

For Hydrogen

Hydrogen pipeline transport is a transportation of hydrogen through a pipe as part of the hydrogen infrastructure. Hydrogen pipeline transport is used to connect the point of hydrogen production or delivery of hydrogen with the point of demand, with transport costs similar to CNG, the technology is proven,. Most hydrogen is produced at the place of demand with every 50 to 100 miles (160 km) an industrial production facility. The 1938-Rhine-Ruhr 240 km hydrogen pipeline is still in operation. As of 2004 there are 900 miles (1,400 km) of low pressure hydrogen pipelines in the USA and 930 miles (1,500 km) in Europe.

For Water

Two millennia ago the ancient Romans made use of large aqueducts to transport water from higher elevations by building the aqueducts in graduated segments that allowed gravity to push the water along until it reached its destination. Hundreds of these were built throughout Europe and elsewhere, and along with flour mills were considered the lifeline of the Roman Empire.

The ancient Chinese also made use of channels and pipe systems for public works. The famous Han Dynasty court eunuch Zhang Rang (d. 189 AD) once ordered the engineer Bi Lan to construct a series of square-pallet chain pumps outside the capital city of Luoyang. These chain pumps serviced the imperial palaces and living quarters of the capital city as the water lifted by the chain pumps was brought in by a stoneware pipe system.

Pipelines are useful for transporting water for drinking or irrigation over long distances when it needs to move over hills, or where canals or channels are poor choices due to considerations of evaporation, pollution, or environmental impact.

The 530 km (330 mi) Goldfields Water Supply Scheme in Western Australia using 750 mm (30 inch) pipe and completed in 1903 was the largest water supply scheme of its time.

Examples of significant water pipelines in South Australia are the Morgan-Whyalla (completed 1944) and Mannum-Adelaide (completed 1955) pipelines.

There are two Los Angeles, California aqueducts, the *First Los Angeles Aqueduct* (completed 1913) and the *Second Los Angeles Aqueduct* (completed 1970) which also include extensive use of pipelines.

For Beverages

For Beer

Bars in the Veltins-Arena, a major football ground in Gelsenkirchen, Germany, are interconnected by a 5 km long beer pipeline. It is the favourite method for distributing beer in such large stadiums, because the bars have to overcome big differences between demands during various stages of a match; this allows them to be supplied by a central tank.

In Randers city in Denmark, the so-called Thor beer pipeline still exists. Originally copper pipes were running directly from the brewery, and, when in 90s the brewery moved out the city, Thor beer replaced the centre of a star with a giant tank.

For Other Uses

The town of Hallstatt in Austria claims to contain "the oldest industrial pipeline in the world", dating back to 1595. It was constructed from 13,000 trunks to transport saline solution for 40 kilometers from Hallstatt to Ebensee.

Types by Transport Function

In general, pipelines can be classified in three categories depending on purpose:

Gathering pipelines : Group of smaller interconnected pipelines forming complex networks with the purpose of bringing crude oil or natural gas from several nearby wells to a treatment plant or processing facility. In this group, pipelines are usually short-a couple of hundred metres-and with small diameters. Also sub-sea pipelines for collecting product from deep water production platforms are considered gathering systems.

Transportation pipelines : Mainly long pipes with large diameters, moving products (oil, gas, refined products) between cities, countries and even continents. These transportation networks include several compressor stations in gas lines or pump stations for crude and multiproducts pipelines.

Distribution pipelines : Composed of several interconnected pipelines with small diameters, used to take the products to the final consumer. Feeder lines to distribute gas to homes and businesses downstream. Pipelines at terminals for distributing products to tanks and storage facilities are included in this group.

Operation

When a pipeline is built, the construction project not only covers the civil work to lay the pipeline and build the pump/compressor stations, it also has to cover all the work related to the installation of the field devices that will support remote operation.

Field devices are instrumentation, data gathering units and communication systems. The field Instrumentation includes flow, pressure and temperature gauges/transmitters, and other devices to measure the relevant data required. These instruments are installed along the pipeline on some specific locations, such as injection or delivery stations, pump stations (liquid pipelines) or compressor stations (gas pipelines), and block valve stations.

The information measured by these field instruments is then gathered in local Remote Terminal Units (RTU) that transfer the field data to a central location in real time using communication systems, such as satellite channels, microwave links, or cellular phone connections.

Pipelines are controlled and operated remotely, from what is usually known as *The Main Control Room*. In this center, all the data related to field measurement is consolidated in one central database. The data is received from multiple RTUs along the pipeline. It is common to find RTUs installed at every station along the pipeline.

The SCADA system at the Main Control Room receives all the field data and presents it to the pipeline operator through a set of screens or Human Machine Interface, showing the operational conditions of the pipeline. The operator can monitor the hydraulic conditions of the line, as well as send operational commands (open/close valves, turn on/off compressors or pumps, change setpoints, etc.) through the SCADA system to the field. To optimize and secure the operation of these assets, some pipeline companies are using what is called *Advanced Pipeline Applications*, which are software tools installed on top of the SCADA system, that provide extended functionality to perform leak detection, leak location, batch tracking (liquid lines), pig tracking, composition tracking, predictive modelling, look ahead modelling, operator training and more.

Technology—Components

Pipeline networks are composed of several pieces of equipment that operate together to move products from location to location. The main elements of a pipeline system are:

Initial injection station : Known also as *supply* or *inlet* station, is the beginning of the system, where the product is injected into the line. Storage facilities, pumps or compressors are usually located at these locations.

Compressor/pump stations : Pumps for liquid pipelines and Compressors for gas pipelines, are located along the line to move the product through the pipeline. The location of these stations is defined by the topography of the terrain, the type of product being transported, or operational conditions of the network.

Partial delivery station : Known also as *intermediate* stations, these facilities allow the pipeline operator to deliver part of the product being transported.

Block valve station : These are the first line of protection for pipelines. With these valves the operator can isolate any segment of the line for maintenance work or isolate a rupture or leak. Block valve stations are usually located every 20 to 30 miles (48 km), depending on the type of pipeline. Even though it is not a design rule, it is a very usual practice in liquid pipelines. The location of these stations depends exclusively on the nature of the product being transported, the trajectory of the pipeline and/or the operational conditions of the line.

Regulator station : This is a special type of valve station, where the operator can release some of the pressure from the line. Regulators are usually located at the downhill side of a peak.

Final delivery station : Known also as outlet stations or terminals, this is where the product will be distributed to the consumer. It could be a tank terminal for liquid pipelines or a connection to a distribution network for gas pipelines.

Leak Detection Systems

Since oil and gas pipelines are an important asset of the economic development of almost any country, it has been required either by government regulations or internal policies to ensure the safety of the assets, and the population and environment where these pipelines run. Pipeline companies face government regulation, environmental constraints and social situations.

Pipeline companies should comply with government regulations which may define minimum staff to run the operation, operator training requirements, up to specifics including pipeline facilities, technology and applications required to ensure operational safety. As an example, in the State of Washington, it is mandatory for pipeline operators to be able to detect and locate leaks of 8 percent of maximum flow within 15 minutes or less.

The social situation also affects the operation of pipelines. In third world countries, product theft is a problem for pipeline companies. It is common to find unauthorized extractions in the middle of the pipeline. In this case, the detection levels should be under 2 percent of maximum flow, with a high expectation for location accuracy.

Different types of technologies and strategies have been implemented, from physically walking the lines to satellite surveillance. The most common technology to protect these lines from occasional leaks is known as Computational Pipeline Monitoring Systems or CPM. CPM takes information from the field related to pressures, flows, and temperatures to estimate the hydraulic behaviour of the product being transported. Once the estimation is done, the results are compared to other field references to detect the presence of an anomaly or unexpected situation, which may be related to a leak.

The American Petroleum Institute has published several articles related to the performance of CPM in liquids pipelines, the API Publications are:

- API 1130 – Computational pipeline monitoring for liquids pipelines
- API 1155 – Evaluation methodology for software based leak detection systems
- API 1149 – Pipeline variable uncertainties & their effects on leak detectability.

Implementation

As a rule pipelines for all uses are laid in most cases underground. However in some cases it is necessary to cross a valley or a river on a pipeline bridge. Pipelines for centralized heating systems are often laid on the ground or overhead. Pipelines for petroleum running through permafrost areas as Trans-Alaska-Pipeline are often run overhead in order to avoid melting the frozen ground by hot petroleum which would result in sinking the pipeline in the ground.

Regulation

In the US, onshore and offshore pipelines used to transport oil and gas are regulated by the Pipeline and Hazardous Materials Safety Administration (PHMSA). Certain offshore pipelines used to produce oil and gas are regulated by the Minerals Management Service (MMS). In Canada, pipelines are regulated by either the provincial regulators or, if they cross provincial boundaries or the Canada/US border, by the National Energy Board (NEB).

Government regulations in Canada and the United States require that buried fuel pipelines must be protected from corrosion. Often, the most economical method of corrosion control is by use of pipeline coating in conjunction with cathodic protection and technology to monitor the pipeline. Above ground, cathodic protection is not an option. The coating is the only external protection.

Pipelines and Geopolitics

Pipelines for major energy resources (petroleum and natural gas) are not merely an element of trade. They connect to issues of geopolitics and international security as well, and the construction, placement, and control of oil and gas pipelines often figure prominently in state interests and actions. A notable example of pipeline politics occurred at the beginning of the year 2009, wherein a dispute between Russia and Ukraine ostensibly over pricing led to a major political crisis. Russian state-owned gas company Gazprom cut off natural gas supplies to Ukraine after talks between it and the Ukrainian government fell through. In addition to cutting off supplies to Ukraine, Russian gas flowing through Ukraine—which included nearly all supplies to Southeastern Europe and some supplies to Central and Western Europe—was cut off, creating a major crisis in several countries heavily dependent on Russian gas as fuel. Russia was accused of using the dispute as leverage in its attempt to keep other powers, and particularly the European Union, from interfering in its "near abroad".

Oil and gas pipelines also figure prominently in the politics of Central Asia and the Caucasus.

Dangers—Accidents

Pipelines conveying flammable or explosive material, such as natural gas or oil, pose special safety concerns.

For a more complete list see List of pipeline accidents:

- 1982-One of the largest non-nuclear explosions in history occurred along the Trans-Siberian Pipeline in the former Soviet Union. It has been alleged that the explosion was the result of CIA sabotage of the Trans-Siberian Pipeline.
- June 4, 1989-sparks from two passing trains detonated gas leaking from an LPG pipeline near Ufa, Russia. Up to 645 people were reported killed.
- October 17, 1998-1998 Jesse pipeline explosion at Jesse in the Niger Delta in Nigeria, a petroleum pipeline exploded killing about 1,200 villagers, some of whom were scavenging gasoline-the worst of several similar incidents in this country.
- June 10, 1999-a pipeline rupture in a Bellingham, Washington park led to the release of 277,200 gallons of gasoline. The gasoline was ignited, causing an explosion that killed two children and one adult.

- August 19, 2000-natural gas pipeline rupture and fire near Carlsbad, New Mexico; this explosion and fire killed 12 members of the same family. The cause was due to severe internal corrosion of the pipeline.
- July 30, 2004-a major natural gas pipeline exploded in Ghislenghien, Belgium near Ath (thirty kilometres southwest of Brussels), killing at least 24 people and leaving 132 wounded, some critically. (CNN) (Expatica)
- May 12, 2006-an oil pipeline ruptured outside Lagos, Nigeria. Up to 200 people may have been killed.
- November 1, 2007-a propane pipeline exploded near Carmichael, Mississippi, about 30 miles (48 km) south of Meridian, Mississippi. Two people were killed instantly and an additional four were injured. Several homes were destroyed and sixty families were displaced. The pipeline is owned by Enterprise Products Partners LP, and runs from Mont Belvieu, Texas, to Apex, North Carolina, according to an Enterprise spokesman.

As Targets

Pipelines can be the target of vandalism, sabotage, or even terrorist attacks. In war, pipelines are often the target of military attacks, as destruction of pipelines can seriously disrupt enemy logistics.

Bibliography

Alex, Marks: *Handbook of Oceanic Pipeline Computations*. Tulsa, OK: PennWell Books, 1980.

Arthur, Cote E. : *Fire Protection Handbook*. Quincy, MA: National Fire Protection Association, 1997.

Benedict, Robert P.: *Fundamentals of Pipe Flow*. New York, NY: Wiley, 1980.

Berry, Mary Clay: *Alaska Pipeline: The Politics of Oil and Native Land Claims*. Bloomington, IN: Indiana University Press, 1975.

Bowie, George L. and Robert Wiegel: *Marine Pipelines: An Annotated Bibliography*. Fort Belvoir, VA: U.S. Army Corps of Engineers, Coastal Engineering Research Center, 1977.

Chen, W.F.: *Civil Engineering Handbook*. Boca Raton, FL: CRC Press, 1995.

Craig, Bruce D. and David S. Anderson: *Handbook of Corrosion Data. Materials Park*, OH: ASM International, 1995.

David Willoughby: *Plastic Piping Handbook*. McGraw-Hill Professional, Delhi, 2002.

DeWald, Omar E.: *Regional Environmental Assessment: Gulf of Mexico Pipeline Activities*. Metaire, LA: U.S. Department of the Interior, The Service, 1983.

Flynn, M.J. : *Computer Architecture: Pipelined and Parallel Processor Design*, Narosa, Delhi, 2011.

Gallick, Edward C.: *Competition in the Natural Gas Pipeline Industry: An Economic Policy Analysis*. Westport, CT: Praeger, 1993.

Guy, N.G.: *Pipe Protection*. New York, NY: Elsevier Applied Science, 1987.

Hosmanek, Max: *Pipeline Construction*. Austin, TX: Petroleum Extension Service, 1984.

James O. Pennock: *Piping Engineering Leadership for Process Plant Projects*, Gulf Professional Publishing, 2001.

Kiefner, J.F. and E.B. Clark: *History of Line Pipe Manufacturing in North America*. New York, NY: American Society of Mechanical Engineers, 1996.

Lamit, Louis Gary: *Pipe Fitting and Piping Handbook*. Englewood Cliffs, NJ: Prentice-Hall, 1984.

Lester, C.B.: *Hydraulics for Pipeliners,* Houston, TX: Gulf Publishing Co., 1994.

Mazurkiewicz, B.K.: *Offshore Platforms and Pipelines: Selected Contributions*. Clausthal-Zellerfeld, West Germany: Trans Tech Publications, 1987.

McKetta, John J.: *Piping Design Handbook*. New York, NY: M. Dekker, 1992.

Mousselli, A.H.: *Offshore Pipeline Design, Analysis, and Methods*. Tulsa, OK: Pennwell Books, 1981.

Nayyar, P.E., Mohinder L.: *Piping Handbook,* New York: McGraw-Hill, 2000.

Pearson, Norman: *Pipelines and Farming*. London, ON, Canada: Strathmore House, 1995.

Peggs, Lawrence A.: *Underground Piping Handbook*. Malabar, FL: R.E. Krieger, 1985.

Raj, Baldev ; B.K. Choudhary and Anish Kumar: *Pressure Vessels and Piping: Materials and Properties*, Narosa, Delhi, 2009.

Roscow, James P.: *The Building of the Alaska Pipeline*. New York, NY: Prentice-Hall, 1977.

Sahu, G.K. *: Handbook Of Piping Design,* New Age International, Delhi, 2009.

Sosnin, H.A.: *Procedures for Pipewelding*. Washington, DC: National Association of Plumbing, 1983.

Thomas Sixsmith & R. Hanselka: *Handbook of Thermoplastic Piping System Design*. Marcel Dekker Ltd, London, 1997.

Thomson, James C.: *Pipejacking and Microtunnelling*. New York, NY: Blackie Academic Professional, 1993.

Tullis, J. Paul: *Hydraulics of Pipelines: Pumps, Valves, Cavitation, Transients*. New York, NY: John Wiley & Sons, 1989.

Vincent-Genod, Jacques: *Fundamentals of Pipeline Engineering*. Houston, TX: Gulf Publishing, 1984.

Warring, R.H.: *Handbook of Valves, Piping, and Pipelines*. Houston, TX: Gulf Publishing Co., 1982.

Watkins, Reynold King and Loren Runar Anderson: *Structural Mechanics of Buried Pipes*. Baton Rouge, CRC Press, 2000.

Williams, Peter J.: *Pipelines and Permafrost: Physical Geography and Development in the Circumpolar North*. New York, NY: Longman, 1979.

Index

❑❑❑